IB Mathematics SL

in **80** pages

2018-2019

This book has been developed independently of the International Baccalaureate Organization (IBO) and the content is in no way connected with nor endorsed by the International Baccalaureate Organization (IBO).

Acknowledgments

The author would like to thank Metaxia, Natasa and Maria for the proofreading. The majority of graphs and diagrams were produced by FX MathPack, a wonderful graphing tool by Efofex Software.

While every attempt has been made to trace and acknowledge copyright, the author apologizes for any accidental infringement where copyright has proved untraceable.

Cover Image: iStock.com/Tramont_ana

Ed. (October 2018)

The best IB Math revision guide app, with video lessons, many calculators, solvers and it's **100% free**, No in-app purchases, No advertisements.

Any comments, improvements, or suggestions can be sent directly to the author at george.feretzakis@mathematics4u.com.

CONTENTS

Introduction

This **updated** revision guide will be a valuable resource and reference for students, assisting them to understand and learn the theory of IB mathematics SL.

The ideal preparation for any student preparing for the IB math exams is to practice doing questions from past papers systematically. Similar exercises can be found in any IB mathematics SL textbook and on official distributor sites for IB materials.

The guide aims to help the IB student by both revising the theory and going through some well-chosen examples of the IB mathematics SL curriculum.

I have made a concerted effort to explain the mathematical terms to the student in a **clear, straightforward** and **understandable manner**. I have aimed to create a **thorough** and **concise** material with an emphasis on **simplicity** which will be effective for both teachers and students.

By presenting the theory that every IB student should know before taking any quiz, test or exam, this revision guide is designed to make the topics of IB mathematics SL both comprehensible and easy to grasp.

George Feretzakis, July 2018

"Truth is ever to be found in the simplicity, and not in the multiplicity and confusion of things."

Isaac Newton

A Get the app

The best IB Math revision guide app, with video lessons, many calculators, solvers and it's **100% free**, **No** in-app purchases, **No** advertisements.

Prior Learning

$(a \pm b)^2 = a^2 \pm 2ab + b^2$		$a^2 - b^2 = (a + b)(a - b)$	
Area of a parallelogram	$A = b \times h$	Volume of a cylinder	$V = \pi r^2 h$
Area of a triangle	$A = \dfrac{1}{2}(b \times h)$	Volume of a cuboid	$V = l \times w \times h$
Area of a trapezium	$A = \dfrac{1}{2}(a + b) \times h$	Area of the curved surface of a cylinder	$A = 2\pi rh$
Area of a circle	$A = \pi r^2$	Volume of a sphere	$V = \dfrac{4}{3}\pi r^3$
Circumference of a circle	$C = 2\pi r$	Volume of a cone	$V = \dfrac{1}{3}\pi r^2 h$

Surface area of a sphere	$A = 4\pi r^2$	Area of the curved surface of a cone	$A = \pi r l$
Volume of a pyramid	$V = \dfrac{1}{3}(Area) \times h$	Volume of a prism	$V = (Area) \times h$

Significant Figure Rules *

There are three rules for determining how many significant figures are in a number:

1. Non-zero digits are always significant.

2. Any zeros between two significant digits are significant.

3. A final zero or trailing zeros in the decimal portion only are significant.

Trailing zeros are a sequence of 0s in the decimal representation of a number, after which no other digits follow.

Example 1	Example 2
1254.04 rounded to 5 s.f. is 1254.0	0.030503062 rounded to 5 s.f. is 0.030503
1254.04 rounded to 4 s.f. is 1254	0.030503062 rounded to 4 s.f. is 0.03050
1254.04 rounded to 3 s.f. is 1250	0.030503062 rounded to 3 s.f. is 0.0305
1254.04 rounded to 2 s.f. is 1300	0.030503062 rounded to 2 s.f. is 0.031
1254.04 rounded to 1 s.f. is 1000	0.030503062 rounded to 1 s.f. is 0.03

* You can download our app using the QR code at page *ii*, where there is a calculator that can round any number to as many significant figures as desired using proper rounding rules. Apart from the aforementioned calculator, our app has many other solvers and calculators accompanied by the corresponding theory.

Sequences, Series and Binomial theorem

A **sequence** is a set of numbers arranged in a definite order.

$$a_1, a_2, a_3, \ldots, a_n, \ldots$$

where $a_1, a_2, a_3, \ldots, a_n, \ldots$ are called **terms**.

Arithmetic Sequence and Series

A sequence $\{a_n\}$ in which each term differs from the previous one by the same constant d (common difference) is called **arithmetic sequence**.

$$a_{n+1} - a_n = d$$

The general term a_n can be found using the formula

$$a_n = a_1 + (n-1)d$$

The Sum (S_n) of the first n terms of an arithmetic sequence $\{a_n\}$ is called **arithmetic series** and given by

$$S_n = \frac{n}{2}(a_1 + a_n)$$

$$S_n = \frac{n}{2}[2a_1 + (n-1)d]$$

Note: If we want to deduce the general term a_n from S_n, we can use the following formula:

$$a_n = S_n - S_{n-1} \text{ and } a_1 = S_1$$

Example

An arithmetic sequence has a 1st term of 80 and a 24th term of 172. Find the 74th term, an expression for the general term and the sum S_n.

Solution

Using the formula $a_n = a_1 + (n-1)d$, we have that

$$a_{24} = a_1 + (24-1)d \Leftrightarrow d = \frac{172 - 80}{23} = \frac{92}{23} = 4$$

Thus, $a_{74} = a_1 + (74-1)d = 80 + 73 \times 4 = 372$

The general term is $a_n = 80 + (n-1)4$

and the sum of the first n terms is $S_n = \frac{n}{2}[160 + (n-1)4]$

Geometric Sequence and Series

A sequence $\{a_n\}$ in which each term can be obtained from the previous one by multiplying a non-zero constant r (common ratio) is called **geometric sequence**.

$$\frac{a_{n+1}}{a_n} = r$$

The general term a_n can be found using the formula

$$a_n = a_1 r^{n-1}$$

The Sum (S_n) of the first n terms of a geometric sequence $\{a_n\}$ is called **geometric series** and given by

$$S_n = \frac{a_1(r^n - 1)}{r - 1}$$

The Sum to infinity (S_∞) of geometric series is given by

$$S_\infty = \frac{a_1}{1 - r}, |r| < 1 \ (-1 < r < 1)$$

If a geometric sequence has a finite sum ($|r| < 1$) then it is called **convergent**. Otherwise, it is called **divergent**.

Example

A geometric sequence has a 4th term of 32 and a 7th term of 256. Find the 10th term, an expression for the general term a_n and the sum S_n.

Solution

Using the formula $a_n = a_1 r^{n-1}$ we have that

$$a_4 = a_1 r^{4-1} \Leftrightarrow 32 = a_1 r^3$$
$$a_7 = a_1 r^{7-1} \Leftrightarrow 256 = a_1 r^6$$

Dividing the two equations we have $\dfrac{a_1 r^3}{a_1 r^6} = \dfrac{32}{256} \Leftrightarrow r^3 = 8 \Leftrightarrow r = 2$

and $32 = a_1 r^3 \Leftrightarrow 32 = a_1 2^3 \Leftrightarrow a_1 = 4$

Thus, $a_{10} = a_1 r^{n-1} = 4 \times 2^{10-1} = 2048$

The general term is $a_n = 4 \times 2^{n-1}$

and the sum $S_n = \dfrac{4(2^n - 1)}{2 - 1} = 4(2^n - 1)$

Sigma Notation

> Sigma notation is a way of expressing sums uses the Greek letter Σ
> $$\sum_{i=1}^{n} a_i = a_1 + a_2 + \cdots + a_n$$
> where i is the index of summation taking values from 1 to n.

Properties of Sigma Notation

$\sum_{i=1}^{n} (a_i \pm b_i) = \sum_{i=1}^{n} a_i \pm \sum_{i=1}^{n} b_i$	$\sum_{i=1}^{n} c a_i = c \sum_{i=1}^{n} a_i$
$\sum_{i=1}^{n} c = cn$	$\sum_{i=k}^{n} a_i = \sum_{i=1}^{n} a_i - \sum_{i=1}^{k-1} a_i$

Binomial Theorem

> $$(a + b)^n = a^n + \binom{n}{1} a^{n-1} b + \ldots + \binom{n}{r} a^{n-r} b^r + \cdots + b^n = \sum_{k=0}^{n} \binom{n}{k} a^{n-k} b^k$$
>
> where $\binom{n}{r} = \dfrac{n!}{r!(n-r)!}$ and $n!$ is the product of the first n positive integers for $n \geq 1$
>
> $$n! = 1 \times 2 \times 3 \times \cdots \times n \text{ and } 0! = 1$$

Properties of the symbol $\binom{n}{r}$

$$\binom{n}{0} = 1 \qquad \binom{n}{1} = n \qquad \binom{n}{n} = 1 \qquad \binom{n}{r} = \binom{n}{n-r}$$

Pascal's Triangle {Blaise Pascal (1623 –1662)}

The Pascal's triangle is a triangular arrangement of numbers that gives the coefficients in the expansion of

any binomial expression, such as $(a + b)^n$.

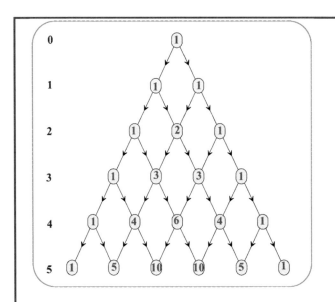

The first row is one 1. Then we have two 1s. The outsides of the triangle are always 1, but the insides are different. To find the number on the next row, add the two numbers above it together. Every row is built from the row above it.

For example the expansion of $(a + b)^3$ according to Pascal's Triangle will be
$$(a + b)^3 = 1a^3 + 3a^2b + 3ab^2 + 1b^3$$

Examples

1.   Using the binomial theorem, expand $(3x + 1)^4$.

Solution

Using the formula

$$(a + b)^n = a^n + \binom{n}{1} a^{n-1}b + \cdots + \binom{n}{r} a^{n-r}b^r + \cdots + b^n$$

we have that

$$(3x + 1)^4 = (3x)^4 + \binom{4}{1}(3x)^{4-1}1 + \binom{4}{2}(3x)^{4-2}1 + \binom{4}{3}(3x)^{4-3}1 + 1^4 =$$

$$= 81x^4 + 108x^3 + 54x^2 + 12x + 1$$

2. Find the term in x^3 in the expansion of $\left(4x - \dfrac{5}{x^2}\right)^6$

Solution

The general term of this binomial expansion is given by the following formula:

$$\binom{6}{r}(4x)^{6-r}\left(\frac{-5}{x^2}\right)^r = \binom{6}{r}4^{6-r}x^{6-r}(-5)^r\left(\frac{1}{x^2}\right)^r = \binom{6}{r}4^{6-r}(-5)^r x^{6-r}(x^{-2})^r =$$

$$= \binom{6}{r}4^{6-r}(-5)^r x^{6-r}x^{-2r} = \binom{6}{r}4^{6-r}(-5)^r x^{6-3r}$$

Since we have to find the term in x^3, we set

$$6 - 3r = 3$$
$$r = 1$$

Therefore the coefficient of x^3 is $\binom{6}{1}4^{6-1}(-5)^1 = 6 \cdot 4^5 \cdot (-5) = -30{,}720$

Note: A useful hint concerning the factorial notation is that, for example, we can express $7!$ as $5! \times 6 \times 7$ instead of $7! = 1 \times 2 \times 3 \times 4 \times 5 \times 6 \times 7$

Compound Interest

The compound interest formula for calculating the **Future Value (FV)** of an amount with a **Present Value (PV)** is:

$$FV = PV\left(1 + \frac{r}{100k}\right)^{kn}$$

where r is the nominal annual interest rate, n the number of years and k is the number of compounding periods per year.

Example

Gregory invests \$80,000 in a savings account earning 5% per year compounded monthly. Calculate the value of the investment after 4 complete years.

Answer

In this example, $r = 5$, $n = 4$ and $k = 12$

$$FV = PV\left(1 + \frac{r}{100k}\right)^{kn} =$$

$$= \$80{,}000\left(1 + \frac{5}{100 \times 12}\right)^{12\times4} = \$97{,}671.63 = \$97{,}700\ (3\ s.f.)$$

Straight Lines

The equation of a straight line is usually written:

$$y = mx + c$$

where m is the **slope** or gradient and c is the **y-intercept**.

Another way to find the equation of a straight line is the following:

$$y - y_0 = m(x - x_0)$$

where m is the slope and (x_0, y_0) is a point which lies on the line.

To find the **slope** m, you use the following formula:

$$m = \frac{y_2 - y_1}{x_2 - x_1} = \frac{rise}{run}$$

where (x_1, y_1) and (x_2, y_2) are two points which lie on the line.

Note: When a line has a **positive** slope it rises left to right (its graph will be **increasing**).

When a line has a **negative** slope, it falls left to right (its graph will be **decreasing**).

The **slope** (or gradient) m describes both the direction and the steepness of the line and is related to its angle of incline θ by the tangent function.

$$m = tan(\theta)$$

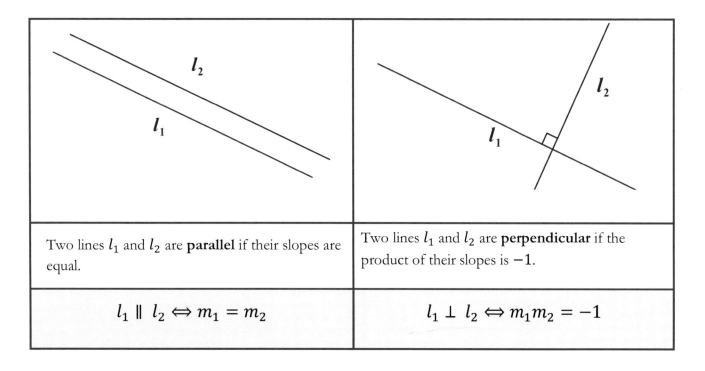

Two lines l_1 and l_2 are **parallel** if their slopes are equal.	Two lines l_1 and l_2 are **perpendicular** if the product of their slopes is -1.
$l_1 \parallel l_2 \Leftrightarrow m_1 = m_2$	$l_1 \perp l_2 \Leftrightarrow m_1 m_2 = -1$

A **horizontal** line graph has an equation of the following form and its **slope is zero**.	A **vertical** line graph has an equation of the following form and it has an **undefined slope** (no slope).
$y = c$	$x = c$

Midpoint Formula: The coordinates of the **midpoint M** of two points $A(x_1, y_1, z_1)$ and $B(x_2, y_2, z_2)$, are given by the following formula:

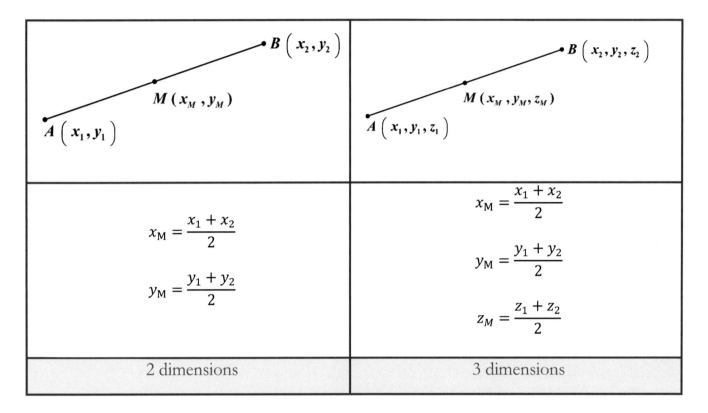

$$x_M = \frac{x_1 + x_2}{2}$$ $$y_M = \frac{y_1 + y_2}{2}$$	$$x_M = \frac{x_1 + x_2}{2}$$ $$y_M = \frac{y_1 + y_2}{2}$$ $$z_M = \frac{z_1 + z_2}{2}$$
2 dimensions	3 dimensions

Distance Formula: Given two points $A(x_1, y_1, z_1)$ and $B(x_2, y_2, z_2)$, the distance (d) between these two points is given by the formula:

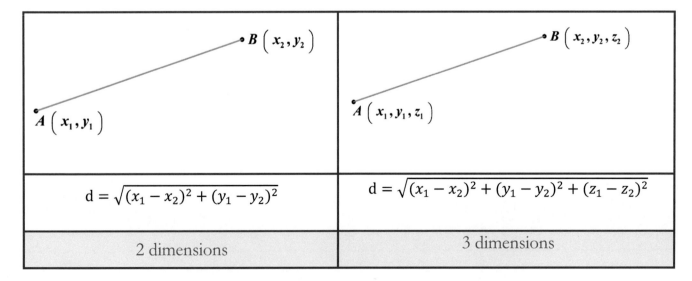

$$d = \sqrt{(x_1 - x_2)^2 + (y_1 - y_2)^2}$$	$$d = \sqrt{(x_1 - x_2)^2 + (y_1 - y_2)^2 + (z_1 - z_2)^2}$$
2 dimensions	3 dimensions

Quadratic Functions

A quadratic function is of the following form

$$f(x) = ax^2 + bx + c, \text{where } a, b, c \in \mathbb{R} \text{ and } a \neq 0.$$

The graph of a quadratic function is a curve called a parabola. Parabolas may open upward or downward, and

all have the same basic "U" shape.

If **a** is **positive**, the graph **opens upward** (Figure 1), and if **a** is **negative** (Figure 2), then it **opens downward**.

The **y-intercept** of the above function is **c**.

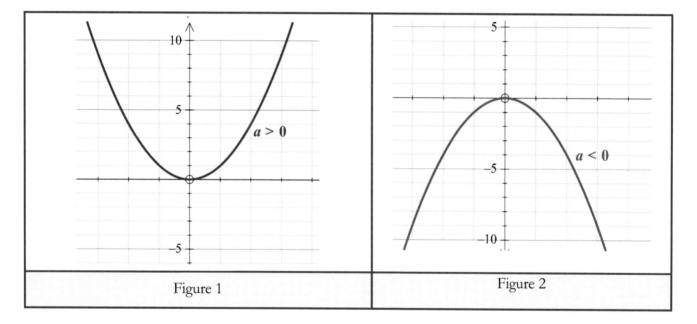

| Figure 1 | Figure 2 |

All parabolas are symmetric with respect to a vertical line called the **axis of symmetry**, with the equation:

$$x = \frac{-b}{2a}$$

A parabola intersects its axis of symmetry at a point called the vertex **V** of the parabola, which has

coordinates:

$$V\left(\frac{-b}{2a}, f\left(\frac{-b}{2a}\right)\right)$$

The three most common forms that are used to express quadratic functions are:

Standard form: $f(x) = ax^2 + bx + c$

Factored form: $f(x) = a(x - r_1)(x - r_2)$, where r_1, r_2 are the two distinct real roots which are the x-intercepts of the graph $f(x)$. The equation of the axis of symmetry (i.e. the x-coordinate of the vertex) passes through the mid-point of the roots of the parabola.

$f(x) = a(x - r)^2$, where r is a real double root (two equal real roots) which is the x-intercept of the graph of $f(x)$.

Vertex form: $f(x) = a(x - h)^2 + k$, where h, k are the coordinates of the vertex $V(h, k)$.

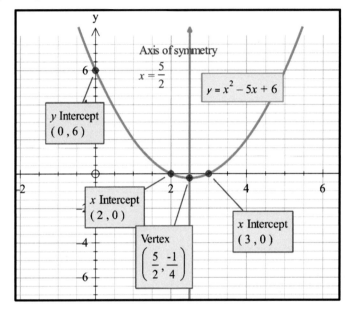

For example, the quadratic function $f(x) = x^2 - 5x + 6$ can be written as

Standard form: $f(x) = x^2 - 5x + 6$

Factored form: $f(x) = 1(x - 2)(x - 3)$, where $2, 3$ are the two roots.

Vertex form: $f(x) = 1(x - \frac{5}{2})^2 - \frac{1}{4}$, where the vertex V has coordinates $(\frac{5}{2}, -\frac{1}{4})$.

Note: We observe that the x-coordinate of the vertex is the midpoint of the two roots, $\frac{2+3}{2} = \frac{5}{2}$.

Quadratic Equations & Inequalities

To **solve a quadratic equation** of the form $ax^2 + bx + c = 0$ follow these steps:

1. When the discriminant $(\Delta = b^2 - 4ac)$ is positive $(\Delta > 0)$ then the equation has two distinct real roots r_1 and r_2.

$$r_{1,2} = \frac{-b \pm \sqrt{b^2 - 4ac}}{2a}$$

2. When the discriminant is equal to zero $(\Delta = 0)$ then the equation has one double (two equal real roots) root r.

$$r = \frac{-b}{2a}$$

3. When the discriminant is negative $(\Delta < 0)$ then the equation has no real roots.

To **solve a quadratic inequality** of the form $ax^2 + bx + c \geq 0$ or $ax^2 + bx + c \leq 0$ follow these steps:

I. When the discriminant is positive $(\Delta > 0)$ then the corresponding quadratic equation $(ax^2 + bx + c = 0)$ has two distinct real roots r_1 and r_2.

In this case the quadratic function $f(x) = ax^2 + bx + c$ has the opposite sign of $\boldsymbol{a}$ between the two real roots and the same sign as $\boldsymbol{a}$ outside the interval (r_1, r_2).

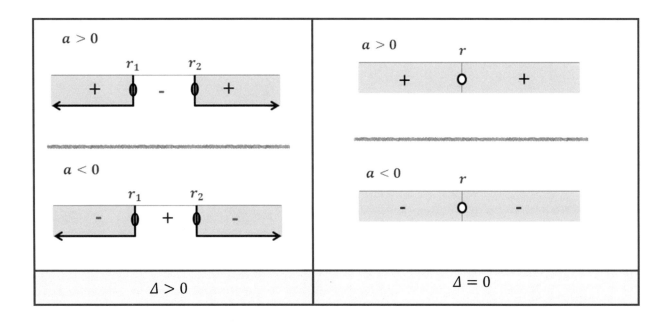

II. When the discriminant is equal to zero $(\varDelta = 0)$ then the corresponding quadratic equation has one double (two equal real roots) real root r. In this case, the quadratic function f has the same sign as a for any value of x except $-\frac{b}{2a}$ since $f\left(-\frac{b}{2a}\right) = 0$.

III. When the discriminant Δ is negative $(\varDelta < 0)$ then the corresponding quadratic equation has no real roots. In this case, the quadratic function f has the same sign as a regardless of the values of x.

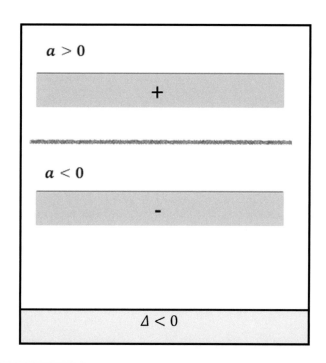

Intersection of a Line and a Parabola

The intersection between a line with equation $y = mx + d$ and a parabola with equation $y = ax^2 + bx + c$ is obtained by setting them equal to each other.

$$ax^2 + bx + c = mx + d \Rightarrow ax^2 + (b - m)x + (c - d) = 0$$

The line intersects the parabola at two points maximum.

- If the discriminant of the quadratic equation above, is negative $(\varDelta < 0)$, then the parabola does not meet the line.
- If the discriminant of the quadratic equation above, is positive $(\varDelta > 0)$, then the line intersects the parabola in two distinct points.
- If the discriminant of the quadratic equation above, is zero $(\varDelta = 0)$, then the line is tangent to the parabola (touches the parabola at only one point).

Functions

Relation: A relation is any set of ordered-pair numbers.

For example, let the relation $R = \{(1,15),(2,17),(3,18),(4,26),(4,67)\}$

The set of all first elements is called the **domain** of the relation.
   The domain of R is the set $\{1,2,3,4\}$.

The set of second elements is called the **range** of the relation.
   The range of R is the set $\{15,17,18,26,67\}$.

Function: A function is a relation in which **no** two ordered pairs have the same first element.

A function associates each element in its domain with **one and only one** element in its range.

All functions are relations, but not all relations are functions.

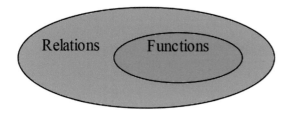

Example

Determine whether the following relations are functions

a) $A = \{(0,5),(3,7),(6,9),(8,15)\}$
b) $B = \{(10,13),(10,31),(20,15),(41,23)\}$

Answer

 a) $A = \{(0,5),(3,7),(6,9),(8,15)\}$ is a function because all the first elements are different.

b) $B = \{(10,13),(10,31),(20,15),(41,23)\}$ is not a function because the first element, 10, is repeated.

The **domain (D_f) of a function** f is the set of all allowable values of the independent variable, commonly known as the x-values.

1. You cannot have negative under a square root or any even radical.
2. You cannot have zero in the denominator.
3. You cannot have a negative number or zero as an argument of a logarithmic function.
4. You cannot have a negative number or zero or one as the base of a logarithmic function.

The **range** (R_f) of a **function** f is the set of y-values when all x-values in the domain are evaluated into the function.

■ A function $f(x)$ is **even** if $f(-x) = f(x)$ for all $x \in D_f$. An **even** function is symmetric with respect to the **y-axis**.

■ A function $f(x)$ is **odd** if $f(-x) = -f(x)$ for all $x \in D_f$. An **odd** function has rotational symmetry with respect to **origin**.

Note: The only function that is both **even** and **odd** is the function $f(x) = 0$.

Vertical Line Test

The vertical line test is a method to determine if a relation is a function. A relation is a function if any vertical line intersects the graph in at most one point.

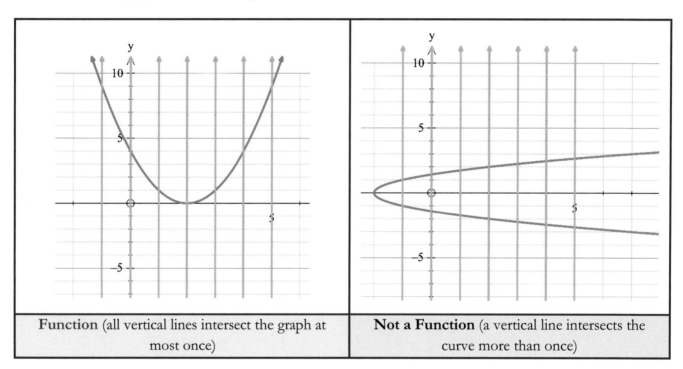

| Function (all vertical lines intersect the graph at most once) | Not a Function (a vertical line intersects the curve more than once) |

Inverse Function

The inverse of the function f is denoted by f^{-1} and is pronounced "f inverse" and <u>it's not</u> the reciprocal of $f\left(f^{-1}(x) \neq \frac{1}{f(x)}\right)$.

To determine algebraically the formula for the inverse of a function $y = f(x)$, you switch y and x to get $x = f(y)$ and then solve for y to get $y = f^{-1}(x)$.

Example

Find the inverse function of $f(x) = \dfrac{2x+3}{x-5}$

Solution

$$y = \frac{2x+3}{x-5}$$ Replace $f(x)$ by y

$$x = \frac{2y+3}{y-5}$$ Switch the x's and y

$$x(y-5) = 2y+3$$ Solve for y

$$xy - 5x = 2y + 3$$

$$xy - 2y = 5x + 3$$

$$y(x-2) = 5x + 3$$

$$y = \frac{5x+3}{x-2}$$

$$f^{-1}(x) = \frac{5x+3}{x-2}$$ Replace y with $f^{-1}(x)$

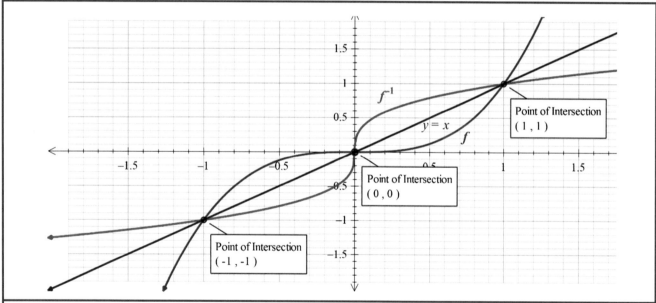

The inverse of a function differs from the function in that all the x-coordinates and y-coordinates have been switched. That is, if for example (4,9) is a point on the graph of the function, then the point (9,4) lies on the graph of the inverse function.

The graph of a function and its inverse are mirror images of each other. They are reflected in the identity function $y = x$.

Important: If the graphs of a function and its inverse intersect at one point then this point will be on the line $y = x$, as shown in the figure above. Therefore, if we want to find the point(s) of intersection between $f(x)$ and $f^{-1}(x)$, instead of finding $f^{-1}(x)$ and then equating both of the functions, we could set $f(x) = x$ and find the common points between $f(x)$ and $y = x$.

Note: A function is said to be a **self-inverse** if $f(x) = f^{-1}(x)$ for all x in the domain.

For example, the reciprocal function $f(x) = \frac{1}{x}$, $(x \neq 0)$ is self-inverse.

Notes: The domain of f^{-1} is equal to the range of f.

The range of f^{-1} is equal to the domain of f.

For example, if $f(4) = 9$ then $f^{-1}(9) = 4$

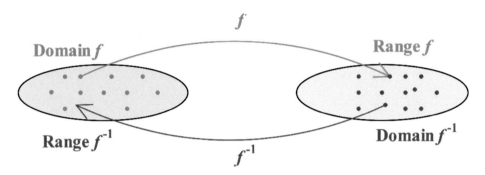

Existence of an Inverse Function

If the function has an inverse that is also a function, then there can only be one y for every x.

A **one-to-one** function is a function in which for every x there is exactly one y and for every y, there is exactly one x. For $f(x)$ to have an **inverse function,** it must be **one-to-one.**

Some functions do not have inverse functions. For example, consider $f(x) = x^2$. This is not an one-to-one function since there are two numbers that f takes to 1, $f(1) = 1$ and $f(-1) = 1$.

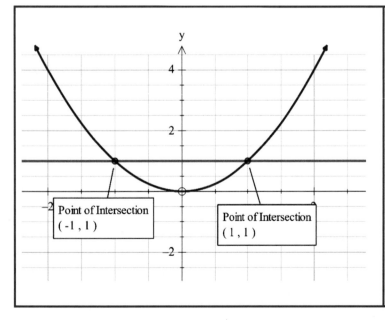

One way to check if a function is one-to-one is the **Horizontal Line Test.**

If a horizontal line intersects the graph of the function more than once, then the function is not one-to-one.

If no horizontal line intersects the graph of the function more than once, then the function is one-to-one.

For example the function $f(x) = x^2$ is not one-to-one since the line $y = 1$ intersects the graph of the function twice.

Composite functions

If we have two functions $f(x)$ and $g(x)$, we can define a composite function

$$(f \circ g)(x) = f(g(x))$$

If a function f has also an inverse then

$$(f \circ f^{-1})(x) = (f^{-1} \circ f)(x) = x$$

where $I(x) = x$ is the identity function.

Example: Given $f(x) = 2x + 3$ and $g(x) = x^2 + 4$, find $f \circ g, g \circ f, f \circ f$ and $g \circ g$.

Solution:

$$(f \circ g)(x) = f\big(g(x)\big) = f(x^2 + 4) = 2(x^2 + 4) + 3$$

$$(g \circ f)(x) = g\big(f(x)\big) = g(2x + 3) = (2x + 3)^2 + 4$$

$$(f \circ f)(x) = f\big(f(x)\big) = f(2x + 3) = 2(2x + 3) + 3$$

$$(g \circ g)(x) = g\big(g(x)\big) = g(x^2 + 4) = (x^2 + 4)^2 + 4$$

Example: The function f is defined by $f(x) = x^3 + 4$.

Find an expression for $g(x)$ in terms of x given that $(f \circ g)(x) = x - 2$.

Solution: If $f(x) = x^3 + 4$ then $f^{-1}(x) = \sqrt[3]{x - 4}$.

If $h(x) = x - 2$ then we have to find a function g such that

$$(f \circ g)(x) = h(x) \Rightarrow \left(\overbrace{f^{-1} \circ f}^{x} \circ g\right)(x) = (f^{-1} \circ h)(x) \Rightarrow$$

$$\Rightarrow (x \circ g)(x) = (f^{-1} \circ h)(x) \Rightarrow g(x) = f^{-1}(h(x)) \Rightarrow g(x) = f^{-1}(x - 2)$$

$$\Rightarrow g(x) = \sqrt[3]{(x - 2) - 4} \Rightarrow g(x) = \sqrt[3]{x - 6}$$

Asymptotes

Horizontal Asymptote

A **Horizontal Asymptote** is a horizontal line that the graph of a function approaches, as x approaches negative or positive infinity. A horizontal asymptote may intersect the graph of the function.

For the following example (Graph 1) the function f has a horizontal asymptote at $y = -1$ and the function g has a horizontal asymptote at $y = 1$.

$$\lim_{x \to \infty} f(x) = -1$$

$$\lim_{x \to -\infty} g(x) = 1$$

Vertical Asymptote

A **Vertical Asymptote** of a curve is a line of the form $x = a$ such that as x approaches some constant value a then the curve goes towards infinity (positive or negative). <u>A vertical asymptote doesn't intersect the graph of the function</u> (Graph 2). For the following example (Graph 2) the function f has a vertical asymptote at $x = 1$ and a horizontal asymptote at $y = 0$.

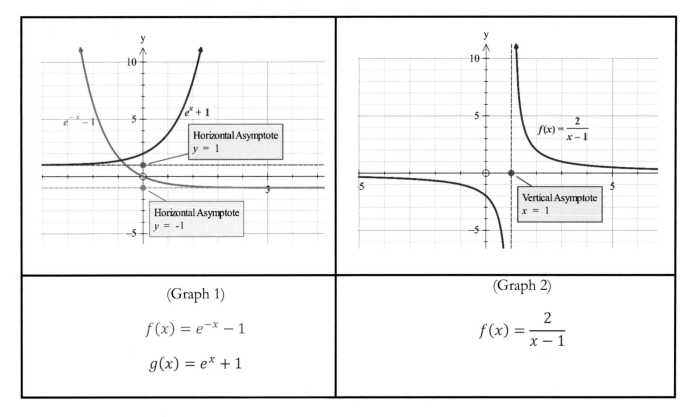

(Graph 1)	(Graph 2)
$f(x) = e^{-x} - 1$	$f(x) = \dfrac{2}{x - 1}$
$g(x) = e^{x} + 1$	

Rational Functions and Asymptotes

Let f be a rational function

$$f(x) = \frac{ax + b}{cx + d} \ , c \neq 0$$

The graph of $y = f(x)$ will have a **vertical asymptote** at this value of x for which the denominator is equal to zero. The graph of $y = f(x)$ may have a horizontal asymptote according to the following:

▪ If the degree of the numerator is equal to the degree of the denominator then the graph of $y = f(x)$ will have a horizontal asymptote at $y = \frac{a}{c}$

▪ If the degree of the numerator is less than the degree of the denominator $(a = 0)$ then the graph of $y = f(x)$ will have a horizontal asymptote at $y = 0$.

Examples

1. The function

$$f(x) = \frac{x - 2}{x + 3}$$

has a **vertical asymptote** at $x = -3$ and a horizontal asymptote at $y = \frac{1}{1} = 1$.

2. The function

$$f(x) = \frac{1}{x + 2}$$

has a **vertical asymptote** at $x = -2$ and a horizontal asymptote at $y = 0$.

3. The function

$$f(x) = \frac{2x - 2}{3x + 1}$$

has a **vertical asymptote** at $x = -\frac{1}{3}$ and a horizontal asymptote at $y = \frac{2}{3}$.

Graph Transformation of Functions

Horizontal Translations

A horizontal translation means that every point (x, y) on the graph of the original function $f(x)$ is transformed to $(x + c, y)$, $(x - c, y)$ on the graph of the transformed function $f(x - c)$ or $f(x + c)$ respectively, where **c** is a **positive** constant.

The graph of $f(x - c)$ is shifted **right** **c** units (Graph 1).

The graph of $f(x + c)$ is shifted **left** **c** units (Graph 2).

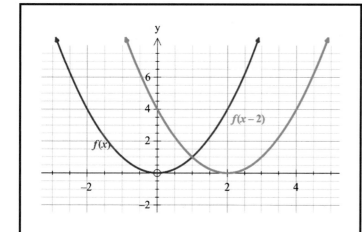	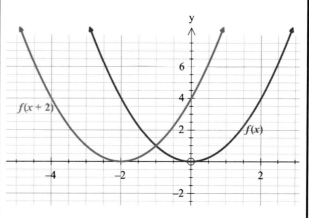
(Graph 1)	(Graph 2)
Horizontal shift 2 units to the **right**	Horizontal shift 2 units to the **left**
$f(x) \rightarrow f(x - 2)$	$f(x + 2) \leftarrow f(x)$
$(0, 0) \rightarrow (2, 0)$	$(-2, 0) \leftarrow (0, 0)$

Vertical Translations

If $f(x)$ is the original function and $c > 0$ then

the graph of $f(x) + c$ is shifted **up c** units (Graph 3)

and the graph of $f(x) - c$ is shifted **down c** units (Graph 4).

A vertical translation means that every point (x, y) on the graph of the original function $f(x)$ is transformed to $(x, y + c)$, $(x, y - c)$ on the graph of the transformed function $f(x) + c$ or $f(x) - c$ respectively.

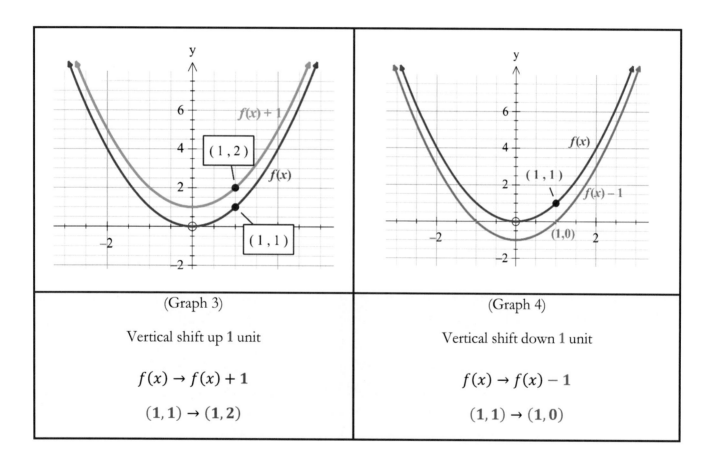

(Graph 3)	(Graph 4)
Vertical shift up 1 unit	Vertical shift down 1 unit
$f(x) \rightarrow f(x) + 1$	$f(x) \rightarrow f(x) - 1$
$(1, 1) \rightarrow (1, 2)$	$(1, 1) \rightarrow (1, 0)$

Note: Sometimes the horizontal and vertical translations are denoted as a vector. For example translation by the vector $\binom{-1}{4}$ denotes a horizontal shift of 1 unit to the left, and vertical shift of 4 units up.

Vertical Stretching and Shrinking (Dilations)

If $f(x)$ is the original function and $a \in \mathbb{R}$ then the graph of $af(x)$ is a **vertical stretch** by **a scale factor of a** (Graph 5).

A vertical stretch means that every point (x, y) on the graph of the original function $f(x)$ is transformed to (x, ay) on the graph of the transformed function $af(x)$.

Horizontal Stretching and Shrinking (Dilations)

If $f(x)$ is the original function and $a \in \mathbb{R}$ then the graph of $f(ax)$ is a **horizontal stretch** by **a scale factor of $\frac{1}{a}$** (Graph 6).

A horizontal stretch means that every point (x, y) on the graph of the original function $f(x)$ is transformed to $(\frac{x}{a}, y)$ on the graph of the transformed function $f(ax)$.

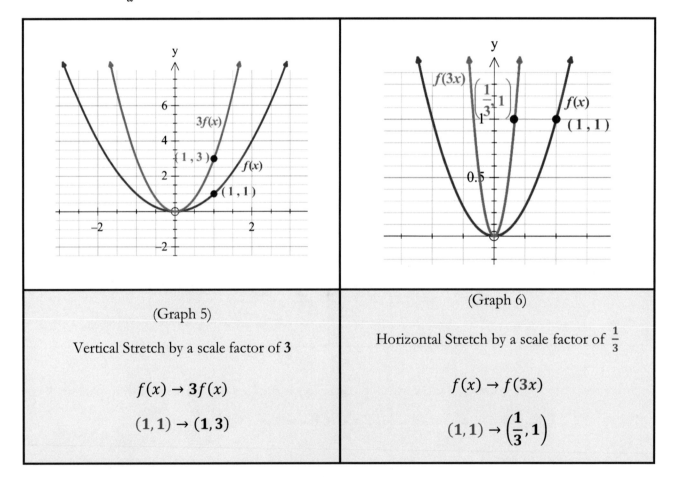

(Graph 5)	(Graph 6)
Vertical Stretch by a scale factor of **3**	Horizontal Stretch by a scale factor of $\frac{1}{3}$
$f(x) \rightarrow 3f(x)$	$f(x) \rightarrow f(3x)$
$(1, 1) \rightarrow (1, 3)$	$(1, 1) \rightarrow \left(\frac{1}{3}, 1\right)$

Reflections

If $f(x)$ is the original function then

the graph of $-f(x)$ is a **reflection in the x-axis** (Graph 7).

and the graph of $f(-x)$ is a **reflection in the y-axis** (Graph 8).

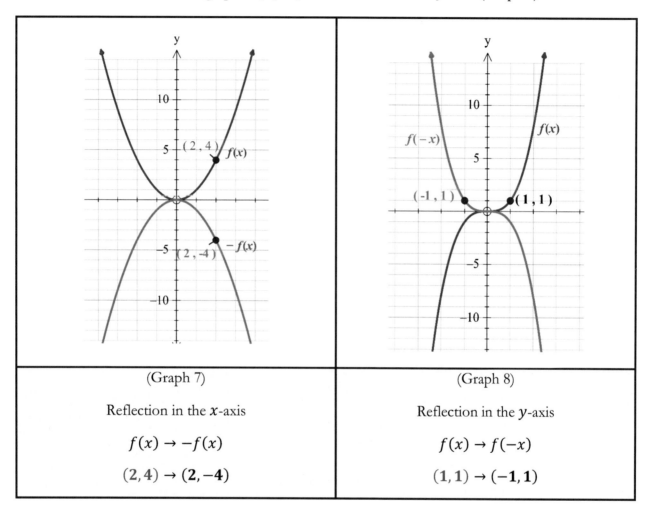

(Graph 7)	(Graph 8)
Reflection in the x-axis	Reflection in the y-axis
$f(x) \to -f(x)$	$f(x) \to f(-x)$
$(2,4) \to (2,-4)$	$(1,1) \to (-1,1)$

Order of Transformations

When we perform multiple transformations the order of these transformations may affect the final graph. Apart from a few exceptions, the intended order could be the following:

1. Horizontal Shifts

2. Stretch / Shrink

3. Reflections

4. Vertical Shifts

Example

How can we obtain the graph of $g(x) = 3\sqrt{2x - 1}$ from the graph of $f(x) = \sqrt{x}$?

Answer

The order of the transformations could be the following:

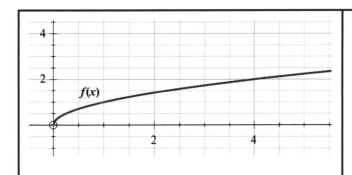

$$f(x) = \sqrt{x}$$

For example, the point $(4,2)$ will be transformed as follows

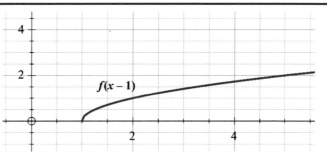

$$f(x - 1) = \sqrt{x - 1}$$

Horizontal Shift 1 unit to the right

$$(4,2) \rightarrow (5,2)$$

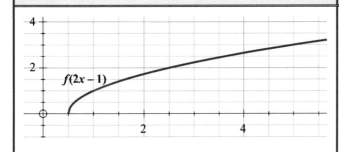

$$f(2x - 1) = \sqrt{2x - 1}$$

Horizontal Shrink by a scale factor of $\frac{1}{2}$

$$(5,2) \rightarrow (\tfrac{5}{2}, 2)$$

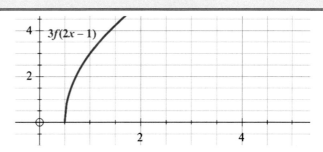

$$3f(2x - 1) = 3\sqrt{2x - 1}$$

Vertical Stretch by a scale factor of 3

$$(\tfrac{5}{2}, 2) \rightarrow (\tfrac{5}{2}, 6)$$

Exponential & Logarithmic Functions

Exponential Function

An Exponential function is a function of the form $f(x) = a^x$ where a is a positive constant and $a \neq 1$.

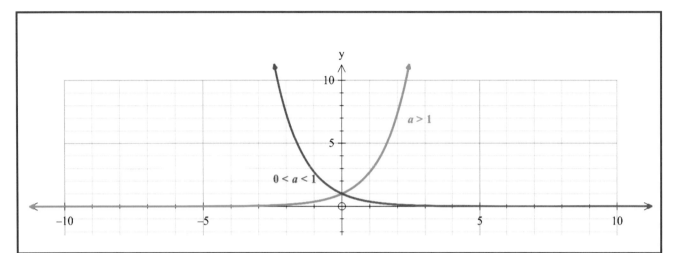

If $a > 1$ then the exponential function increases.	If $0 < a < 1$ then the exponential function decreases.
In either case, the x-axis is its horizontal asymptote	
The domain of $f(x) = a^x$ consists of all real numbers and its range consists of positive numbers only.	

Laws of Exponents

If a, b are positive numbers, x, y are real numbers and m, n are positive integers then:

$a^{x+y} = a^x a^y$	$\left(\dfrac{a}{b}\right)^x = \dfrac{a^x}{b^x}$	$a^{\frac{m}{n}} = \sqrt[n]{a^m}$
$a^{x-y} = \dfrac{a^x}{a^y}$	$a^0 = 1$	$a^{\frac{1}{n}} = \sqrt[n]{a}$
$(a^x)^y = a^{xy}$	$a^{-x} = \dfrac{1}{a^x}$	
$(ab)^x = a^x b^x$	$a^x = \dfrac{1}{a^{-x}}$	

Note: To solve exponential equations without using logarithms, you need to have equations with the same base to some other power

For example: $2^{x+3} = 2^{3x+5}$ then $x + 3 = 3x + 5 \Rightarrow 2x = -2 \Rightarrow x = -1$.

Logarithmic Function

The inverse of the exponential function $f(x) = a^x$ is the logarithmic function with base a

$$f^{-1}(x) = \log_a x, \text{ where } a, x > 0 \text{ and } a \neq 1$$

If $0 < a < 1$ then the logarithmic function decreases.

If $a > 1$ then the logarithmic function increases.

In either case, the y-axis is its **vertical asymptote**.

Domain: $x > 0$
Range: $y \in \mathbb{R}$

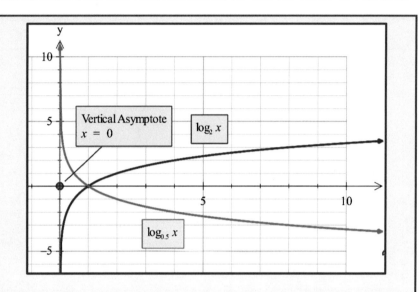

The graph of
$$f(x) = \log_a x$$
can be obtained by reflecting the graph of
$$g(x) = a^x$$
across the line $y = x$ since they **are inverse to each other**.

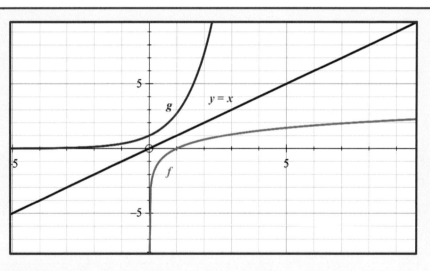

$$\log_a x = y \iff a^y = x$$

$$\log_a a^x = x \text{ for every } x \in \mathbb{R}$$

$$a^{\log_a x} = x \text{ for every } x > 0$$

$$\log_a 1 = 0$$

$$\log_a a = 1$$

> The logarithm with base e is referred to as a **natural logarithm** $log_e x \equiv lnx$ where e is the [1]Euler's number which is defined as $e = \lim_{n \to \infty} \left(1 + \frac{1}{n}\right)^n = 2.71828..$
>
> [1]{Leonhard Euler (1707 –1783)}

Laws of Logarithms

If a, b are positive numbers, $a, b \neq 1$ and x, y are real positive numbers, and k is a real number then:

- $log_\alpha(xy) = log_\alpha x + log_\alpha y$

- $log_\alpha \left(\frac{x}{y}\right) = log_\alpha x - log_\alpha y$

- $log_\alpha \left(\frac{1}{y}\right) = -log_\alpha y$

- $log_\alpha x^k = k log_\alpha x$

- $log_\alpha x = \frac{log_b x}{log_b a}$ (change of base)

Solving Logarithmic equations containing only logarithms

If an equation contains only two logarithms, on opposite sides of the equal sign, with the same base and both with a coefficient of one, then we can drop the logarithms.

$$log_\alpha x = log_\alpha y$$

then

$$x = y$$

Important: Use only solutions that are in the intersection of the domains of the logarithmic terms. We need to make sure that when we plug our solutions back into the original equation, we get a positive number. Otherwise, we must reject these solutions.

Example

$$log_3(x + 3) = log_3(2x - 5)$$

$$x + 3 = 2x - 5$$

$$x = 8$$

We plug this solution back into the original equation an, we get a positive number, so it is accepted.

Solving Logarithmic Equations Containing Terms without Logarithms

We simplify the problem using the properties of logarithms and then rewrite the logarithmic problem in exponential form.

Example Solve the logarithmic equation
$$log_3(x + 3) = 4$$

Solution

$$log_3(x + 3) = 4$$
$$x + 3 = 3^4$$
$$x = 81 - 3 = 78$$

Solving Exponential Equations using Logarithms

We simplify the problem using the properties of logarithms and then rewrite the exponential problem in logarithmic form.

Example Solve the exponential equation
$$2^{x-3} = 15$$

Solution

$$2^{x-3} = 15$$

$$log_2 2^{x-3} = log_2 15$$
$$(x - 3)log_2 2 = log_2 15$$
$$(x - 3)1 = log_2 15$$
$$x = log_2 15 + 3$$

Example The number of bacteria in a colony, B, is modeled by the function $B(t) = 500 \times 3^{0.1t}$ where t is measured in days.
(a) Find the initial number of bacteria in this colony.
(b) Find the number of bacteria after 10 days.
(c) How long does it take for the number of bacteria in the colony to reach 2000?

Solution

(a) $B(0) = 500 \times 3^{0.1 \times 0} = 500$
(b) $B(10) = 500 \times 3^{0.1 \times 10} = 500 \times 3 = 1500$
(c) $B(t) = 500 \times 3^{0.1t} = 2000 \Rightarrow 3^{0.1t} = 4 \Rightarrow 0.1t ln3 = ln4 \Rightarrow t = \dfrac{ln4}{0.1 ln3} = 12.6$ days

Trigonometry

◌ **Degrees - Radians measurement of angles**

$$2\pi \ radians = 360^o \ degrees$$

◌ **Circle Sectors and Segments**

A **sector** of a circle is a region of a circle bounded by a central angle and its intercepted arc.
The region of a circle bounded by an arc and a chord is called a **segment** of a circle.

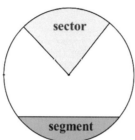

Arc length L	Sector Area
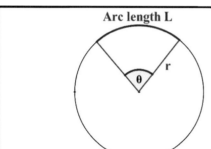	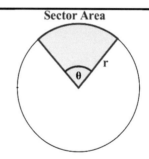
Arc Length: $L = \theta \, r$ θ: in radians	**Sector Area:** $A = \dfrac{1}{2} \, \theta \, r^2$ θ: in radians
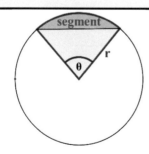	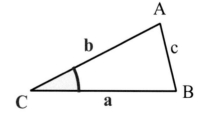
Area of segment = **Area of the sector - Area of the triangle=** $\dfrac{1}{2} \, \theta \, r^2 - \dfrac{1}{2} \, r^2 \sin\theta$	**Two sides and the included angle** **Area of a triangle:** $A = \dfrac{1}{2} \, a \, b \, \sin C$

Trigonometric ratios

For the following **right-angled triangle** we have

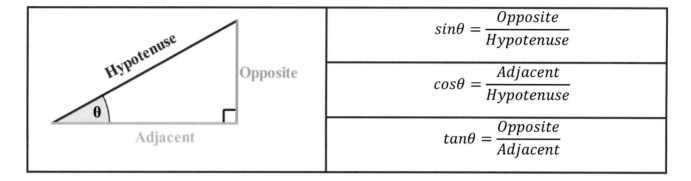

$$sin\theta = \frac{Opposite}{Hypotenuse}$$

$$cos\theta = \frac{Adjacent}{Hypotenuse}$$

$$tan\theta = \frac{Opposite}{Adjacent}$$

The Unit Circle is the circle with center (0,0) and radius 1 unit

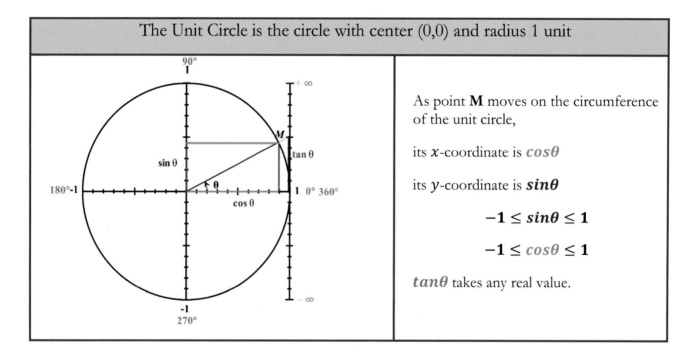

As point **M** moves on the circumference of the unit circle,

its x-coordinate is $cos\theta$

its y-coordinate is $sin\theta$

$$-1 \leq sin\theta \leq 1$$

$$-1 \leq cos\theta \leq 1$$

$tan\theta$ takes any real value.

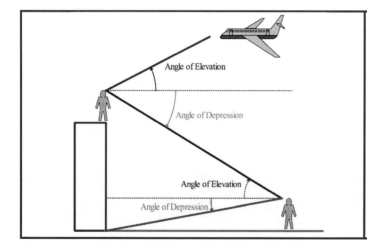

The **angle of elevation** denotes the angle from the horizontal upward to a point.

The **angle of depression** denotes the angle from the horizontal downward to a point.

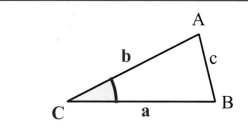

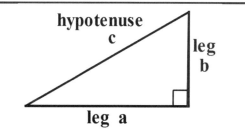

| Two sides and the included angle

Area of a triangle:

$$A = \frac{1}{2}\,a\,b\,sinC$$ | [1]Pythagoras' Theorem

In a right-angled triangle with legs a, b and hypotenuse c, it holds

$$c^2 = a^2 + b^2$$ |

[1]{Pythagoras of Samos(570 – 495 BC)}

Trigonometric table of common angles

Degrees	0° or 360°	30°	45°	60°	90°	180°	270°
Radians	0 or 2π	$\frac{\pi}{6}$	$\frac{\pi}{4}$	$\frac{\pi}{3}$	$\frac{\pi}{2}$	π	$\frac{3\pi}{2}$
sin	0	$\frac{1}{2}$	$\frac{\sqrt{2}}{2}$	$\frac{\sqrt{3}}{2}$	1	0	-1
cos	1	$\frac{\sqrt{3}}{2}$	$\frac{\sqrt{2}}{2}$	$\frac{1}{2}$	0	-1	0
tan	0	$\frac{\sqrt{3}}{3}$	1	$\sqrt{3}$	Is not defined	0	Is not defined

Trigonometric table of related angles

	$\frac{\pi}{2}-\theta$	$\frac{\pi}{2}+\theta$	$\pi-\theta$	$\pi+\theta$	$\frac{3\pi}{2}-\theta$	$\frac{3\pi}{2}+\theta$	$-\theta$
sin	$cos\theta$	$cos\theta$*	$sin\theta$	$-sin\theta$	$-cos\theta$	$-cos\theta$	$-sin\theta$
cos	$sin\theta$	$-sin\theta$	$-cos\theta$	$-cos\theta$	$-sin\theta$	$sin\theta$	$cos\theta$

* For example $sin\left(\frac{\pi}{2}+\theta\right) = cos\theta$

Trigonometric Identities

$sin^2\theta + cos^2\theta = 1$	$tan\theta = \dfrac{sin\theta}{cos\theta}$
$sin2\theta = 2sin\theta cos\theta$	$cos2\theta = cos^2\theta - sin^2\theta = 2cos^2\theta - 1 = 1 - 2sin^2\theta$

▣ Example of a trigonometric equation
Find the exact solutions of $sin2x = cosx$ for $0 \leq x \leq 2\pi$

$$sin2x = cosx$$

$$2sinx\, cosx = cosx$$

$$2sinx\, cosx - cosx = 0$$

$$cosx\,(2sinx - 1) = 0$$

$$cosx = 0 \qquad\qquad \text{or} \qquad\qquad 2sinx - 1 = 0$$

$$x = \frac{\pi}{2} \text{ or } x = \frac{3\pi}{2} \qquad\qquad\qquad\qquad sinx = \frac{1}{2}$$

$$x = \frac{\pi}{6} \text{ or } x = \frac{5\pi}{6}$$

Important: Trigonometric equations can be solved both graphically and analytically depending on the kind of question and whether this question requires GDC or not.

Example: Given that $sinx = \frac{1}{3}$, where $\frac{\pi}{2} \leq x \leq \pi$, evaluate $sin2x$.

Solution

$$sin^2x + cos^2x = 1$$

$$\left(\frac{1}{3}\right)^2 + cos^2x = 1$$

$$cos^2x = 1 - \frac{1}{9}$$

$$cosx = \pm\sqrt{\frac{8}{9}} \text{ and since } \frac{\pi}{2} \leq x \leq \pi, \text{ therefore } cosx = -\sqrt{\frac{8}{9}}$$

Therefore, $sin2x = 2sinx\, cosx = 2\,\frac{1}{3}\left(-\sqrt{\frac{8}{9}}\right) = -\frac{4\sqrt{2}}{9}$

Sine rule

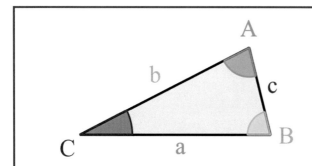

In any triangle ABC:

$$\frac{a}{\sin A} = \frac{b}{\sin B} = \frac{c}{\sin C}$$

__Ambiguous case:__ When you are given two adjacent sides of a triangle followed by an angle, the **Sine rule** will actually give you two answers. The $sin^{-1}x$ function will only give us one of the two angles if we are using a calculator. To find the other one, we need to subtract the calculator's answer from 180° or π.

Cosine rule

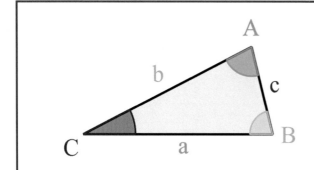

In any triangle ABC
$$a^2 = b^2 + c^2 - 2\,b\,c\cos A$$
$$b^2 = c^2 + a^2 - 2\,c\,a\cos B$$
$$c^2 = a^2 + b^2 - 2\,a\,b\cos C$$

$$\cos A = \frac{b^2 + c^2 - a^2}{2bc} \qquad \cos B = \frac{c^2 + a^2 - b^2}{2ca} \qquad \cos C = \frac{a^2 + b^2 - c^2}{2ab}$$

The **bearing** to a point is the angle measured in a **clockwise** direction from the **north**.

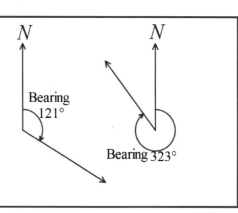

Trigonometric Functions

$f(x) = sinx$

Domain: $x \in \mathbb{R}$

Range: $-1 \leq y \leq 1$

Period: 2π

The **Period** is the length that it takes for the curve to start repeating itself.
The **amplitude** of a trigonometric functions $(\sin x, \cos x)$ is the distance between the principal axis and one of the maximum or minimum points.

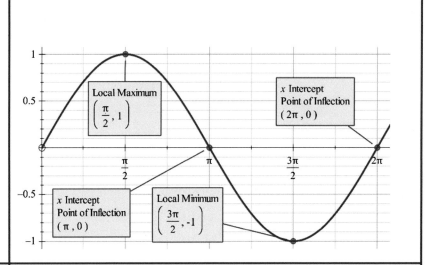

Amplitude: $\frac{y_{max}-y_{min}}{2} = \frac{1-(-1)}{2} = 1$

Principal axis: $y = \frac{y_{max}+y_{min}}{2} = \frac{1+(-1)}{2} = 0$ (the x-axis)

$f(x) = cosx$

Domain: $x \in \mathbb{R}$

Range: $-1 \leq y \leq 1$

Period: 2π

The distance between the x-coordinates of the minimum and maximum of the graphs of $sinx$ or $cosx$ is half of a period.

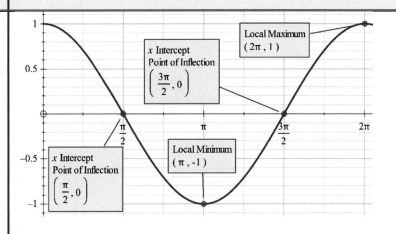

Amplitude: $\frac{y_{max}-y_{min}}{2} = \frac{1-(-1)}{2} = 1$

Principal axis: $y = \frac{y_{max}+y_{min}}{2} = \frac{1+(-1)}{2} = 0$ (the x-axis)

$f(x) = tanx$

Domain: $x \in \mathbb{R}\backslash\left\{\pm\frac{\pi}{2}, \pm\frac{3\pi}{2}, \pm\frac{5\pi}{2}, ...\right\}$

Range: $y \in \mathbb{R}$

Period: π

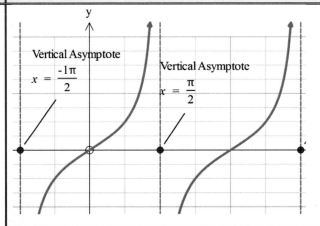

Vertical Asymptotes: $x = \frac{\pi}{2}, x = -\frac{\pi}{2}, x = \frac{3\pi}{2}, ...$

Trigonometric Graph Transformations

$a\,sin[b\,(x-h)]+v$ $a\,cos[b\,(x-h)]+v$	$a\,tan[b\,(x-h)]+v$
Amplitude: $\|a\|$, **Period:** $\frac{2\pi}{b}$, h: **Horizontal shift** v: **Vertical shift, principal axis:** $y=v$	a: **Vertical stretch, Period:** $\frac{\pi}{b}$, h: **Horizontal shift** v: **Vertical shift** The tangent function <u>does not</u> have an amplitude because it does not have any minimum or maximum value.

Example

Find the period, the amplitude, the principal axis, the horizontal and the vertical shift of the following trigonometric functions.

(i) $2sin(3x)+3$

(ii) $4cos(2(x-4))-2$

(iii) $-\frac{1}{2}sin(8(x+2))+5$

(iv) $3tan(4(x+1))$

Solution

(i) Period: $\frac{2\pi}{3}$, amplitude: 2, principal axis: $y=3$, vertical shift: 3 units upward.

(ii) Period: $\frac{2\pi}{2}=\pi$, amplitude: 4, principal axis: $y=-2$, horizontal shift: 4 units to the right, vertical shift: 2 units downward.

(iii) Period: $\frac{2\pi}{8}=\frac{\pi}{4}$, amplitude: $\frac{1}{2}$, principal axis: $y=5$, horizontal shift: 2 units to the left.

(iv) Period: $\frac{\pi}{4}$, horizontal shift: 1 unit to the left and a vertical stretch by a scale factor of 3.

Important: Interesting applications of trigonometric functions are the height of tide and the motion of a Ferris wheel.

Note: There are a lot of applications of sine and cosine rules on problems include navigation, three-dimensional shapes, and angles of elevation and depression.

Differentiation

Derivative

The **tangent line** to the curve $y = f(x)$ at the point $A(x, f(x))$ is the line through A with slope (gradient)

$$f'(x) = \lim_{h \to 0}\left(\frac{f(x+h) - f(x)}{h}\right)$$

which is called the **derivative** of $f(x)$ at x.

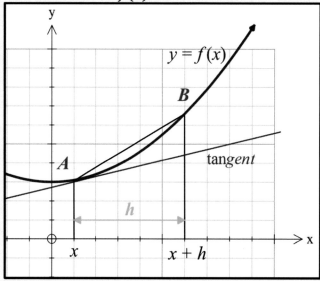

The **rate of change** of the function f at $A(x, f(x))$ is given by the gradient of the tangent to the curve at A.

Apart from the [1]**Newtonian notation (prime)** there is also the [2]**Leibniz notation** for the derivative of the function $y = f(x)$

$$f'(x) = y' = \frac{dy}{dx} = \frac{df}{dx} = \frac{d}{dx}f(x)$$

$\dfrac{dy}{dx}$ measures the **rate of change** of y in respect of x.

Higher Derivatives

$$f'(x) = \frac{dy}{dx}, f''(x) = \frac{d^2y}{dx^2}, f'''(x) = \frac{d^3y}{dx^3}, ..., f^{(n)}(x) = \frac{d^ny}{dx^n}$$

[1]{Sir Isaac Newton (1643-1727)}, [2]{Gottfried Wilhelm Leibniz (1646-1716)}

Example

Find, from first principles, the derivative of $f(x) = x^2$.

Solution

At any point x the derivative is given by

$$f'(x) = \lim_{h \to 0} \frac{f(x+h) - f(x)}{h} = \lim_{h \to 0} \frac{(x+h)^2 - x^2}{h} =$$

$$= \lim_{h \to 0} \frac{x^2 + 2xh + h^2 - x^2}{h} = \lim_{h \to 0} \frac{(2x+h)h}{h} = \lim_{h \to 0}(2x+h) = 2x$$

Thus, the derivative (or gradient) function is $f'(x) = 2x$

Differentiation Rules

▢ $\left(f(x) \pm g(x)\right)' = f'(x) \pm g'(x)$

▢ $\left(cf(x)\right)' = cf'(x)$

▢ $\left(f(x) \times g(x)\right)' = f'(x) \times g(x) + f(x) \times g'(x)$ **Product rule (Newtonian notation)**

▢ $y = uv \Rightarrow \dfrac{dy}{dx} = u\dfrac{dv}{dx} + v\dfrac{du}{dx}$ **Product rule (Leibniz notation)**

▢ $\left(\dfrac{f(x)}{g(x)}\right)' = \dfrac{f'(x) \cdot g(x) - f(x) \cdot g'(x)}{(g(x))^2}$ **Quotient rule (Newtonian notation)**

▢ $y = \dfrac{u}{v} \Rightarrow \dfrac{dy}{dx} = \dfrac{v\frac{du}{dx} - u\frac{dv}{dx}}{v^2}$ **Quotient rule (Newtonian notation)**

▢ $(f \circ g)'(x) = f'\left(g(x)\right) \times g'(x)$ **Chain rule**

▢ $y = g(u), u = f(x) \Rightarrow \dfrac{dy}{dx} = \dfrac{dy}{du} \times \dfrac{du}{dx}$ **Chain rule (Leibniz notation)**

Function	Derivative
$f(x) = c$	$f'(x) = 0$
$f(x) = x$	$f'(x) = 1$
$f(x) = x^k, \quad k \in \mathbb{R}$	$f'(x) = kx^{k-1}, \quad k \in \mathbb{R}$
$f(x) = tanx$	$f'(x) = \dfrac{1}{cos^2x}$
$f(x) = sinx$	$f'(x) = cosx$
$f(x) = cosx$	$f'(x) = -sinx$
$f(x) = e^x$	$f'(x) = e^x$
$f(x) = lnx\ , x > 0$	$f'(x) = \dfrac{1}{x}, x > 0$
$f(x) = \dfrac{1}{x}\ , x \neq 0$	$f'(x) = -\dfrac{1}{x^2}, x \neq 0$
Composite Function	Derivative
$f(x) = [g(x)]^k$	$f'(x) = k[g(x)]^{k-1}g'(x)$
$f(x) = sin(g(x))$	$f'(x) = cos(g(x)) \times g'(x)$
$f(x) = cos(g(x))$	$f'(x) = -sin(g(x)) \times g'(x)$
$f(x) = e^{g(x)}$	$f'(x) = e^{g(x)} \times g'(x)$
$f(x) = \ln[g(x)]\ , g(x) > 0$	$f'(x) = \dfrac{g'(x)}{g(x)}$

Examples

1. Find the derivative of $f(x) = \frac{x^2 e^x + sin^3 x}{lnx}$

Solution (Newtonian notation)

$$f'(x) = \left(\frac{x^2 e^x + sin^3 x}{lnx}\right)' = \frac{(x^2 e^x + sin^3 x)' lnx - (x^2 e^x + sin^3 x)(lnx)'}{(lnx)^2} =$$

$$= \frac{[(x^2 e^x)' + (sin^3 x)'] lnx - (x^2 e^x + sin^3 x)\frac{1}{x}}{(lnx)^2} =$$

$$= \frac{2xe^x lnx + x^2 e^x lnx + 3sin^2 x \, cosx \, lnx - (x^2 e^x + sin^3 x)\frac{1}{x}}{(lnx)^2}$$

2. Find the derivative of $y = (4x - 12)^6$

Solution (Leibniz notation)

Let $u = 4x - 12$ and $y = u^6$

Thus, $\frac{dy}{du} = 6u^5$ and $\frac{du}{dx} = 4$

Since $\frac{dy}{dx} = \frac{dy}{du}\frac{du}{dx} \Rightarrow \frac{dy}{dx} = 6u^5 \cdot 4 = 24u^5 = 24(4x - 12)^5$

Equations of Tangent and Normal

The equation of the **tangent** to a curve $y = f(x)$ at (x_1, y_1) is given by

$$y - y_1 = f'(x_1)(x - x_1)$$

Since the slope (m_T) of the tangent is $m_T = f'(x_1)$.

The equation of the **normal** (the perpendicular to the tangent) to a curve $y = f(x)$ at (x_1, y_1) is given by

$$y - y_1 = -\frac{1}{f'(x_1)}(x - x_1)$$

Since the slope (m_N) of the normal is $m_N = -\frac{1}{m_T} = -\frac{1}{f'(x_1)}$

Note: When the first derivative (the slope of the tangent) at a certain point (x_1, y_1) is equal to zero $(f'(x_1) = 0)$ then the equation of the tangent at this point is given by the equation: $y = y_1$ (a horizontal line) and the equation of the normal at this point is a vertical line of the form $x = x_1$.

Example

Find the equations of the tangent and the normal to the curve $y = x^3$ at the point $(2,8)$.

Solution

$$f'(x) = 3x^2$$

$$f'(2) = 3 \times 2^2 = 12$$

So, the slope of the tangent is $m_T = 12$, the slope of the normal is $m_N = -\frac{1}{12}$

and the corresponding equations are given by the following formulas:

The equation of the **tangent** at $(2,8)$: $y - 8 = 12(x - 2)$

The equation of the **normal** at $(2,8)$: $y - 8 = -\frac{1}{12}(x - 2)$

Stationary points

A **stationary point** is a point where $f'(x) = 0$. It could be a local **minimum**, local **maximum** or a **stationary point of inflection**.

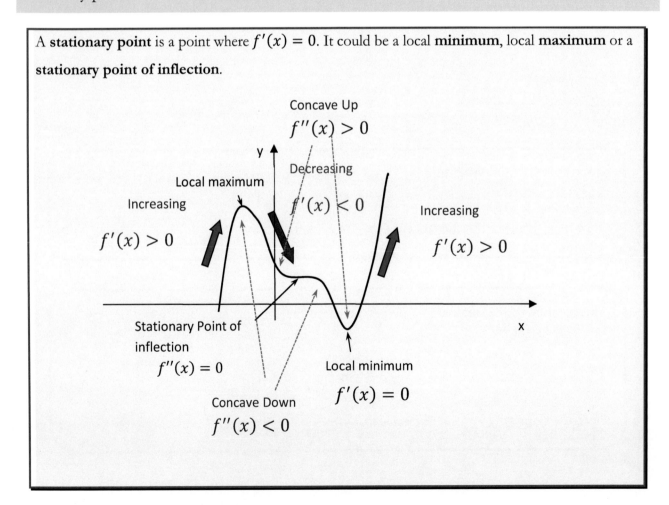

Points of inflection (inflexion)

A **point of inflection** is a point on a curve at which a change of **concavity** occurs.

We have a point of inflection at $x = x_0$ if $f''(x_0) = 0$ **and** the sign of $f''(x)$ changes on either side of $x = x_0$.

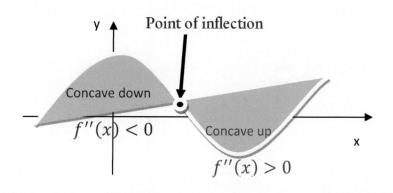

Test for monotonicity (increase / decrease)

- If $f'(x) > 0$ on an interval, then f is **increasing** on this interval.

- If $f'(x) < 0$ on an interval, then f is **decreasing** on this interval.

Test for concavity (concave-up / concave-down)

- f is **concave-up** on an interval I if $f''(x) > 0$ for all x on I.

- f is **concave-down** on an interval I if $f''(x) < 0$ for all x on I.

The First Derivative Test for turning points (maximum /minimum)

Suppose x_0 is a stationary point $(f'(x_0) = 0)$ of a continuous function f.

- If f' changes from positive to negative at x_0, then f has a **local maximum** at x_0.

		x_0	
f'	+	0	−
f	↗	max	↘

■ If f' changes from negative to positive at x_0, then f has a **local minimum** at x_0.

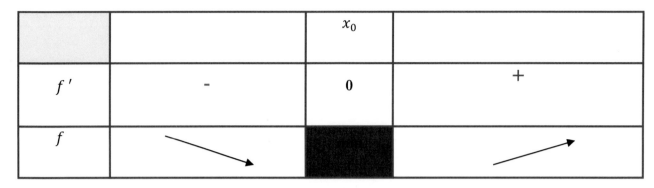

The Second Derivative Test for turning points (maximum/minimum)

▩ If $f'(x_0) = 0$ and $f''(x_0) > 0$, then f has a local minimum at x_0.

▩ If $f'(x_0) = 0$ and $f''(x_0) < 0$, then f has a local maximum at x_0.

Note: If we have a closed interval, apart from the stationary points we should also examine the endpoints for maximum or minimum. This is a common thing when we try to find a minimum or a maximum in an optimization problem.

The graph of the derivative function $f'(x)$

The following guidelines are useful in order to sketch the derivative function $f'(x)$ given the graph of the original function $f(x)$ and vice versa:

1. If the graph of $f(x)$ is **increasing** then $f'(x)$ is **positive** and vice versa.

2. If the graph of $f(x)$ is **decreasing** then $f'(x)$ is **negative** and vice versa.

3. If the graph of $f(x)$ is **concave-up** then the graph of $f'(x)$ is **increasing** and vice versa.

4. If the graph of $f(x)$ is **concave-down** then the graph of $f'(x)$ is **decreasing** and vice versa.

5. If the graph of $f(x)$ has a **stationary point** then the graph of $f'(x)$ has **a zero** and vice versa.

6. If the graph of $f(x)$ has a **point of inflection** then the graph of $f'(x)$ has **a turning point** and vice versa.

Example

We can obtain the graph of $f'(x)$ from the graph of $f(x)$ by applying the rules above

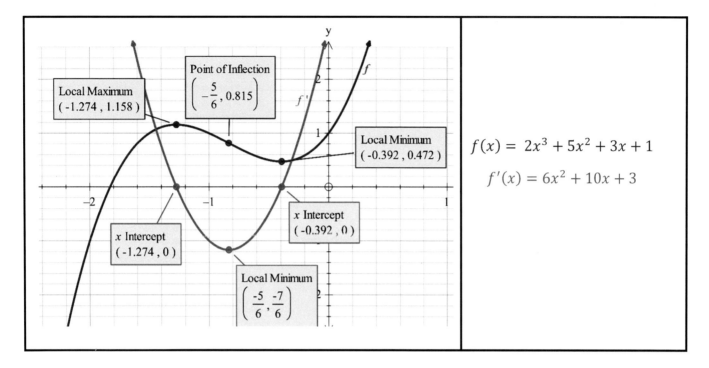

Local Maximum
(-1.274 , 1.158)

Point of Inflection
$\left(-\dfrac{5}{6}, 0.815 \right)$

Local Minimum
(-0.392 , 0.472)

x Intercept
(-1.274 , 0)

x Intercept
(-0.392 , 0)

Local Minimum
$\left(\dfrac{-5}{6}, \dfrac{-7}{6} \right)$

$f(x) = 2x^3 + 5x^2 + 3x + 1$

$f'(x) = 6x^2 + 10x + 3$

Optimization Problems

Many application problems in calculus involve functions for which you want to find maximum or minimum values. First, we have to write down the "constraint" equation and the "optimization" equation. Then, we have to express the optimization equation as a function of only one variable, and, if it is needed, reduce it to be easily differentiable. Finally we have to differentiate the function and set it equals zero in order to find the stationary points. If we have a closed interval, apart from the stationary points we should also examine the endpoints for maximum or minimum. This is a common thing when we try to find a minimum or a maximum in an optimization problem.

Example

A closed cylindrical tin is to be made from a sheet of metal measuring $400 \, cm^2$. Find the dimensions of the tin if the volume is to be maximum.

Solution

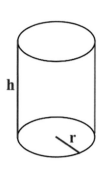

The total surface area of a closed cylinder is

$$A = 2\pi r^2 + 2\pi rh = 400$$

$$h = \frac{400 - 2\pi r^2}{2\pi r}$$

and the volume is $V = \pi r^2 h = \pi r^2 \frac{400 - 2\pi r^2}{2\pi r} = 200r - \pi r^3$

$$V'(r) = (200r - \pi r^3)' = 200 - 3\pi r^2 = 0$$

$$r = \sqrt{\frac{200}{3\pi}}$$

Since $V'(r) > 0$ when $r < \sqrt{\frac{200}{3\pi}}$ and $V'(r) < 0$ when $r > \sqrt{\frac{200}{3\pi}}$, we can conclude

that the volume is maximized when $r = \sqrt{\frac{200}{3\pi}}$ and $h = \dfrac{400 - 2\pi \sqrt{\frac{200}{3\pi}}^2}{2\pi \sqrt{\frac{200}{3\pi}}}$

Curve sketching of $y = f(x)$

- Identify the **domain** of f, that is, the set of values of x for which f is defined.

- Find f' and f''

- Find x -axis **intercepts** setting $y = 0$ and solve for x.

- Find y -axis **intercept** setting $x = 0$ and solve for y.

- Find **Horizontal** (behavior of f as $x \to \pm\infty$) and **Vertical Asymptotes** (where the function is not defined).

- Find where the curve is **increasing** ($f'(x) > 0$) and where it is **decreasing** ($f'(x) < 0$).

- Find where the curve is **concave up** ($f''(x) > 0$) and where it is **concave down** ($f''(x) < 0$).

- Find **Local Minimum, Maximum** values and **points of inflection.**

  Optional Find any **Symmetry** (<u>even function</u>: $f(-x) = f(x)$, which indicates symmetry about y-axis, <u>odd function</u>: $f(-x) = -f(x)$, which indicates symmetry about the origin) the curve may have.

Example

Sketch the graph of $f(x) = \frac{x}{x^2-1}$

Solution

■ The domain is $\{x | x \neq -1,1\}$.

■ The x-axis and y-axis intercept is the origin $(0,0)$.

■ $f'(x) = \frac{(x)'(x^2-1)-x(x^2-1)'}{(x^2-1)^2} = \frac{(x^2-1)-2x^2}{(x^2-1)^2} = -\frac{x^2+1}{(x^2-1)^2} < 0$

■ $f''(x) = \left(-\frac{x^2+1}{(x^2-1)^2}\right)' = \cdots = \frac{2x(x^2+3)}{(x^2-1)^3}$

	$-\infty$	-1	0	1	$+\infty$
$f''(x)$	-	+	-	+	
$f'(x)$	-	-	-	-	
$f(x)$	Decreasing and Concave down	Decreasing and Concave up	Decreasing and Concave down	Decreasing and Concave up	

■ The function has vertical asymptotes at $x = 1$ and $x = -1$ and a horizontal asymptote $y = 0$.

■ There is no local minimum or maximum although we have an inflection point at $(0,0)$.

■ The function is odd because $f(-x) = \frac{-x}{(-x)^2-1} = -f(x)$ which means that there is symmetry about the origin.

Finally the graph of $f(x) = \frac{x}{x^2-1}$ is the following:

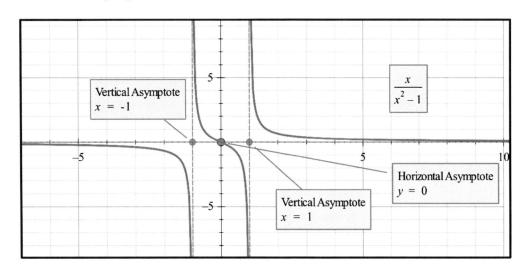

Integration

Indefinite Integral

If $F(x)$ is a function where $F'(x) = f(x)$ then the antiderivative of $f(x)$ is $F(x)$

and the indefinite integral is defined as

$$\int f(x)dx = F(x) + c$$

Definite Integral

A definite integral is of the form

$$\int_{a}^{b} f(x)dx$$

where x is called the variable of integration and $\boldsymbol{a}, \boldsymbol{b}$ are called the **Lower** and **Upper limit**
respectively.

$$\int_{a}^{b} f(x)d\,x = F(b) - F(a) = [F(x)]_{a}^{b}$$

where $F(x)$ is the antiderivative of $f(x)$

Properties of Indefinite Integrals

- $\int f'(x)dx = f(x) + c$

- $(\int f(x)dx)' = f(x)$

- $\int kf(x)dx = k \int f(x)dx$, where k is any constant.

- $\int(f(x) \pm g(x))dx = \int f(x)dx \pm \int g(x)dx$

Properties of Definite Integrals

- $\int_a^a f(x)dx = 0$

- $\int_a^b f(x)dx = -\int_b^a f(x)dx$

- $\int_a^b (-f(x))dx = -\int_a^b f(x)dx$

- $\int_a^b kf(x)dx = k\int_a^b f(x)dx$, where k is any constant

- $\int_a^b f(x)dx + \int_b^c f(x)dx = \int_a^c f(x)dx$

- $\int_a^b ((f(x) \pm g(x))dx = \int_a^b f(x)dx \pm \int_a^b g(x)dx$

Basic Indefinite Integrals

$\int k\,dx = kx + c$	
$\int x^n dx = \dfrac{x^{n+1}}{n+1} + c$ $(n \neq -1)$	$\int (ax+b)^n dx = \dfrac{1}{a}\dfrac{(ax+b)^{n+1}}{n+1} + c$ $(n \neq -1)$
$\int \dfrac{1}{x}dx = \ln x + c,\ x > 0$	$\int \dfrac{1}{ax+b}dx = \dfrac{1}{a}\ln(ax+b) + c$ $,\ ax+b > 0$
$\int e^x dx = e^x + c$	$\int e^{ax+b}dx = \dfrac{1}{a}e^{ax+b} + c$
$\int \sin x\,dx = -\cos x + c$	$\int \sin(ax+b)\,dx = -\dfrac{1}{a}\cos(ax+b) + c$
$\int \cos x\,dx = \sin x + c$	$\int \cos(ax+b)\,dx = \dfrac{1}{a}\sin(ax+b) + c$

Integration Techniques

Integration by Substitution for indefinite integrals

$$\int f(g(x))g'(x)\,dx = \int f(u)du$$

where $u = g(x)$ and $du = g'(x)dx$

Example

Using the substitution $u = x^2 + 5$ find $\int x(x^2 + 5)^4\,dx$

Solution

If $u = x^2 + 5$ then $\frac{du}{dx} = 2x \implies du = 2xdx$

Therefore, $\int x(x^2 + 5)^4\,dx = \frac{1}{2}\int (x^2 + 5)^4\,2xdx = \frac{1}{2}\int u^4\,du =$

$$= \frac{1}{2}\frac{u^5}{5} + c = \frac{(x^2 + 5)^5}{10} + c$$

Integration by Substitution for definite integrals

$$\int_a^b f(g(x))g'(x)\,dx = \int_{g(a)}^{g(b)} f(u)du$$

where $u = g(x)$ and $du = g'(x)dx$

Example

Using the substitution $u = x^4 + 2018$ find

$$\int_4^5 \frac{x^3}{x^4 + 2018}\,dx$$

Solution

If $u = x^4 + 2018$ then $\frac{du}{dx} = 4x^3 \implies du = 4x^3 dx$

To find the new limits of integration we note that

when $x = 4$, $u = 4^4 + 2018 = 256 + 2018 = 2274$

and when $x = 5$, $u = 5^4 + 2018 = 625 + 2018 = 2643$

Therefore

$$\int_4^5 \frac{x^3}{x^4 + 2018} dx = \frac{1}{4} \int_4^5 \frac{4x^3}{x^4 + 2018} dx = \frac{1}{4} \int_{2274}^{2643} \frac{1}{u} du = \frac{1}{4} [\ln u]_{2274}^{2643} =$$

$$= \frac{1}{4} (\ln 2643 - \ln 2274)$$

Splitting the numerator

In order to evaluate an integral with a fractional expression, it is sometimes helpful to split the numerator so as to produce a simpler integral.

Examples

▪ $\int \frac{x+4}{x+10} dx = \int \frac{x+10-6}{x+10} dx = \int \frac{x+10}{x+10} dx - \int \frac{6}{x+10} dx = x - 6\ln(x + 10) + c$

(We assume that $x + 10 > 0$)

▪ $\int \frac{x}{3x+10} dx = \frac{1}{3} \int \frac{3x}{3x+10} dx = \frac{1}{3} \int \frac{3x+10-10}{3x+10} dx = \frac{1}{3} \int \frac{3x+10}{3x+10} dx - \frac{1}{3} \int \frac{10}{3x+10} dx =$

$$= \frac{1}{3} \int dx - \frac{1}{3} \int \frac{10}{3x + 10} dx = \frac{1}{3} x - \frac{10}{9} \ln(3x + 10) + c$$

(We assume that $3x + 10 > 0$)

Applications of Integration

Areas above and below the x-axis

For a function $f(x) \geq 0$ on an interval $[a, b]$, the area between the x-axis and the curve $y = f(x)$ between $x = a$ and $x = b$, is given by

$$A_1 = \int_a^b f(x)dx$$

For a function $f(x) \leq 0$ on an interval $[b, c]$, the area between the x-axis and the curve $y = f(x)$ between $x = b$ and $x = c$, is given by

$$A_2 = -\int_b^c f(x)dx$$

Finally, the total area between the curve and the x-axis on the interval $[a, c]$ is given by

$$A = A_1 + A_2$$

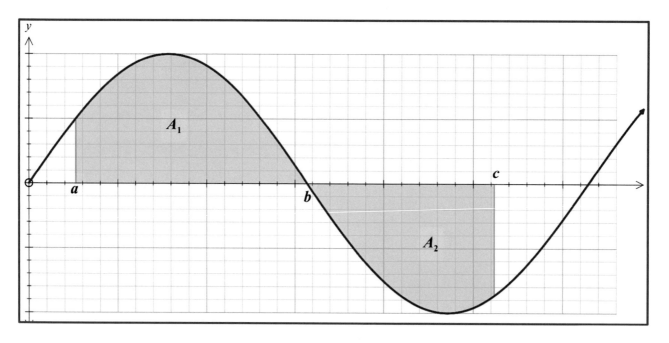

Example

Find the area of the region enclosed by the curve $f(x) = x(x-1)(x-2)$ and the x-axis.

Solution

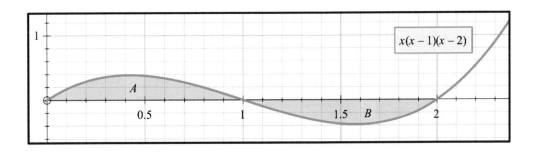

The curve crosses the x-axis at $x = 0, x = 1$ and $x = 2$.

The function is positive on the interval $[0,1]$ and negative on $[1,2]$.

Therefore the required area is

$$A + B = \int_0^1 x(x-1)(x-2)\,dx + \left| \int_1^2 x(x-1)(x-2)\,dx \right| =$$

$$= \int_0^1 (x^3 - 3x^2 + 2x)\,dx - \int_1^2 (x^3 - 3x^2 + 2x)\,dx =$$

$$= \left[\frac{1}{4}x^4 - x^3 + x^2 \right]_0^1 - \left[\frac{1}{4}x^4 - x^3 + x^2 \right]_1^2 =$$

$$= \frac{1}{4} - 1 + 1 - \left(4 - 8 + 4 - \frac{1}{4} + 1 - 1 \right) = \frac{1}{2} \text{ square units}$$

Area between two curves

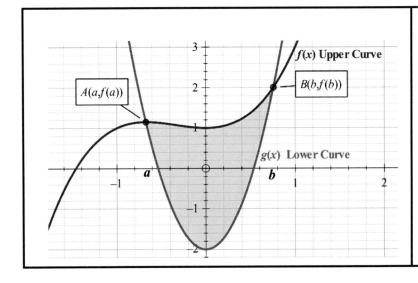

The area between two curves where $f(x) \geq g(x)$ on an interval $[a, b]$, is given by

$$A = \int_a^b (f(x) - g(x))\,dx$$

Example

Find the area of the region enclosed by the curves $f(x) = x^2$ and $g(x) = x^3$.

Solution

To find the coordinates of the common points solve together $x^2 = x^3 \Rightarrow$

$\Rightarrow x^3 - x^2 = 0 \Rightarrow x^2(x - 1) = 0 \Rightarrow x = 0$ or $x = 1$

Solving the inequality $x^2 \geq x^3$ or sketching the graph on GDC, we observe that $f(x) \geq g(x)$ on the interval $[0,1]$.

Therefore the shaded area is given by

$A = \int_0^1 (x^2 - x^3)dx = \left[\frac{1}{3}x^3 - \frac{1}{4}x^4 \right]_0^1 = \frac{1}{3} - \frac{1}{4} = \frac{1}{12}$ square units

Volumes of revolution

When a plane region bounded by the curve $y = f(x)$ and the vertical lines $x = a$ and $x = b$ is revolved about the x-axis, the volume of revolution is given by

$$V_x = \pi \int_a^b y^2 dx$$

When a plane region enclosed by the curves $y = f(x)$ (upper curve) and $y = g(x)$ (lower curve), and the vertical lines $x = a$ and $x = b$ is revolved about the x-axis, the volume of revolution, is given by

$$V_x = \pi \int_a^b ([f(x)]^2 - [g(x)]^2)dx$$

Example

Find the volume of the solid generated when the line $y = 2x$ for $0 \leq x \leq 2$ is revolved around the x- axis.

Solution

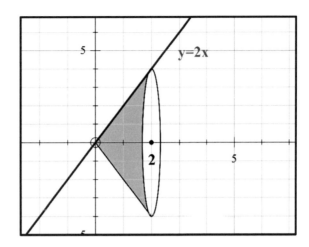

The volume of the revolution around the x- axis is given by

$$V_x = \pi \int_0^2 y^2 dx = \pi \int_0^2 4x^2 dx = \pi \left[\frac{4}{3}x^3\right]_0^2 = \frac{32\pi}{3} \; cubic \; units$$

Kinematics

Let a particle P moves in a straight line and its displacement (position) is $s(t)$ relative to a point O at a given time t.

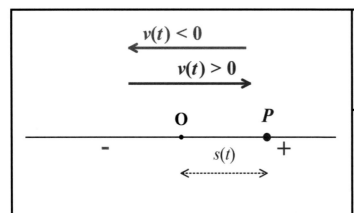

	If $s(t) = 0$ then P is at O.
	If $s(t) > 0$ then P is to the right of O.
	If $s(t) < 0$ then P is to the left of O.
	If $v(t) = 0$ then P is at rest.
	If $v(t) > 0$ then P is moving to the right.
	If $v(t) < 0$ then P is moving to the left.

If $a(t) = 0$ then the velocity may be a minimum or maximum or constant.

If $a(t) > 0$ then the velocity is increasing.

If $a(t) < 0$ then the velocity is decreasing.

The **velocity** $v(t)$ of the particle P at time t is the rate of change of the displacement.

$$v(t) = \frac{ds}{dt}$$

The **acceleration** $a(t)$ of the particle P at time t is the rate of change of the velocity or the second derivative of the displacement.

$$a(t) = \frac{dv}{dt} = \frac{d^2s}{dt^2}$$

The **speed** of the particle P at time t is the absolute value of the velocity at this time.

$$Speed = |v(t)|$$

Important: If the signs of $v(t)$ and $a(t)$ are both positive or both negative (they have the same sign) then the speed of the particle is increasing. If the signs of $v(t)$ and $a(t)$ are opposite then the speed of the particle is decreasing.

The **velocity** $v(t)$ of the particle P is the integral of the acceleration.

$$v(t) = \int a(t)dt$$

The **displacement** $s(t)$ of the particle P is the integral of the velocity.

$$s(t) = \int v(t)dt$$

The **total distance** (d) travelled from t_1 to t_2 is given by the following definite integral:

$$d = \int_{t_1}^{t_2} |v(t)|dt$$

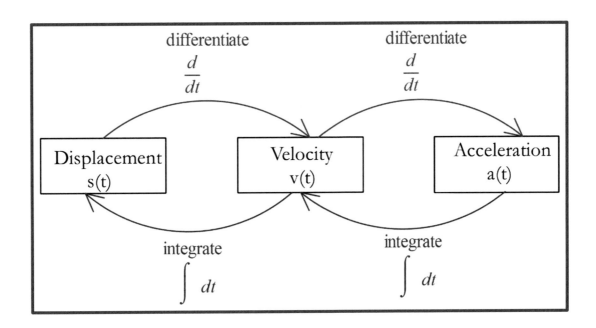

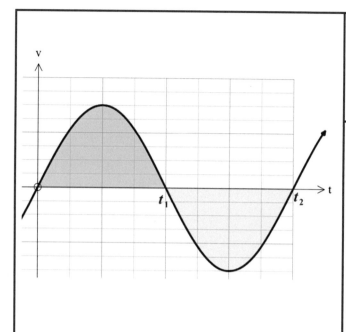

The **displacement** over the interval $[0, t_2]$ is given by

$$s = \int_0^{t_2} v(t)dt$$

The **distance** covered over the interval $[0, t_2]$ is given by

$$d = \int_0^{t_1} v(t)dt - \int_{t_1}^{t_2} v(t)dt$$

$$= \int_0^{t_2} |v(t)|dt$$

Example

The velocity v of a particle at time t is given by $v(t) = t^3 + 2t^2 + 10t + 4$ and that when $t = 4\ sec$ the displacement $s = 14\ m$ find the formula for the displacement at any time t.

Solution

$$s = \int (t^3 + 2t^2 + 10t + 4)dt = \frac{1}{4}t^4 + \frac{2}{3}t^3 + \frac{10}{2}t^2 + 4t + c$$

When $t = 4$ then $s = 14$

$$\frac{1}{4}4^4 + \frac{2}{3}4^3 + \frac{10}{2}4^2 + 4 \cdot 4 + c = 14 \Rightarrow c = -\frac{566}{3}$$

Therefore the displacement is given by $s(t) = \frac{1}{4}t^4 + \frac{2}{3}t^3 + \frac{10}{2}t^2 + 4t - \frac{566}{3}$.

Statistics

Population: The entire group you want to know something about.

Sample: The group you use to infer something about the population.

Random Sample is a set of n objects in a population of N objects where all possible samples are equally likely to happen.

Continuous data can be assigned an infinite number of values between whole numbers. (A person's height or weight).

Discrete data is data that can be counted. For example, the number of students.

A **Bar chart** is a graph that uses vertical or horizontal bars to represent the frequencies of the categories in a data set. (Categorical variables)

The **Histogram** is a graphical display of a frequency or a relative frequency distribution that uses classes and vertical bars of various heights to represent the frequencies. (Quantitative variables)

A **frequency polygon** is a graph that displays the data using lines to connect points plotted for the frequencies. The frequencies represent the heights of the vertical bars in the histograms.

A **stem and leaf plot** is a table where all the data must first sort in ascending order and then each data value is split into a "stem" (the first digit or digits) and a "leaf" (usually the last digit).

Measures of Central Tendency

Mean of a set of values is the number obtained by adding the values and dividing the total by the number of values. In the formula below x_i represents an observation of the data, f_i the corresponding frequency and n the total number of observations.

$$\bar{x} = \frac{1}{n}\sum_{i=1}^{k} f_i x_i = \frac{f_1 x_1 + f_2 x_2 + \cdots + f_k x_k}{n}$$

$$n = \sum_{i=1}^{k} f_i = f_1 + f_2 + \cdots + f_k$$

The Population mean is usually denoted by μ and the sample mean is denoted by $\bar{x}$ (read as 'x-bar').

Notes:

- Calculation of mean can be performed by using the formula or technology.
- When the data are grouped into classes, we should use the midpoint or mid-interval value to represent all values within that interval in order to estimate the mean of grouped data.

Example

Find the mean of a sample of 6 test grades (80, 65, 91, 75, 82, 76)

Answer

$$\bar{x} = \frac{\sum_{i=1}^{6} x_i}{6} = \frac{80 + 65 + 91 + 75 + 82 + 76}{6} \cong 78.17$$

Example (Grouped Data)

Estimate the mean and write down the modal class of the following heights

Heights (cm)	158-160	161-163	164-166	167-169	170-172	173-175	176-178	179-181
Mid-height x_i(cm)	159	162	165	168	171	174	177	180
Frequency f_i	2	4	4	5	7	6	3	1

From the table above, we have that

$\sum_{i=1}^{8} f_i = 2 + 4 + 4 + 5 + 7 + 6 + 3 + 1 = 32$

and $\sum_{i=1}^{8} f_i x_i = 159 \times 2 + 162 \times 4 + 165 \times 4 + 168 \times 5 + 171 \times 7 + 174 \times 6 + 177 \times 3 + 180 \times 1 = 5418$

$$\bar{x} = \frac{\sum f_i x_i}{\sum f_i} = \frac{5418}{32} = 169.31$$

The **modal class** is 170-172 since has the largest frequency (7).

- The **Median** of a data set is the middle value when the data values are arranged in ascending or descending order. If the data set has an even number of entries, the median is the mean of the two middle data entries.

Examples

1. Sample (8, 3, 5, 12, 15, 20, 1), n=7

Answer

Step 1: arrange in ascending order   1, 3, 5, **8**, 12, 15, 20

Step 2: the median is 8

2. Sample (8, 3, 5, 12, 15, 20, 1, 13), n=8

Answer

Step 1: arrange in ascending order   1, 3, 5, **8**, **12**, 13, 15, 20

Step 2: the median (m) is given by $m = \frac{8+12}{2} = 10$

- The **Mode** of a data set is the value that occurs most frequently. When two values occur with the same greatest frequency, each one is a mode and the data set is bimodal. When more than two values occur with the same greatest frequency, each is a mode and data set is said to be multimodal. When no value is repeated, we say there is no mode.

Example

Find the mode of a sample of 7 test grades (80, 65, 91, 75, 82, 76, 82).

Answer

The **mode** is the grade 82 because it has the greatest frequency (=2).

- In a **negatively skewed distribution**, the mean is to the left of the median, and the mode is to the right of the median.

Mean<Median<Mode

- In a **symmetrical distribution**, the data values are evenly distributed on both sides of the mean. Also, when the distribution is unimodal, the mean, median and mode are all equal to one another and are located at the center of the distribution.

Mean=Median

- In a **positively skewed distribution**, the mean is to the right of the median, and the mode is to the left of the median.

Mode<Median<Mean

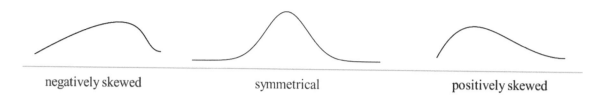

| negatively skewed | symmetrical | positively skewed |

Measures of Dispersion

Range=largest measurement-smallest measurement

Example

Sample (12, 15, 18, 20, 17)

Range=max-min=20-12=8

Variance	Standard Deviation
$s_n^2 = \dfrac{1}{n}\displaystyle\sum_{i=1}^{n}(x_i - \bar{x})^2$	$s_n = \sqrt{\dfrac{1}{n}\displaystyle\sum_{i=1}^{n}(x_i - \bar{x})^2}$

Notes:

- Standard deviation is the square root of the variance.
- Calculation of standard deviation and variance using only technology.

Percentiles

Let $0 < p < 100$. The pth percentile is a number x such that $p\%$ of all measurements fall below the pth percentile and $(100 - p)\%$ fall above it.

Lower Quartile (25th percentile)	Median (50th percentile)	Upper Quartile (75th percentile)
Q_1	Q_2	Q_3

The Interquartile range (IQR) of a data set is the difference between the third and first quartile.

$$\text{IQR}=Q_3 - Q_1$$

Box-and-whisker plot

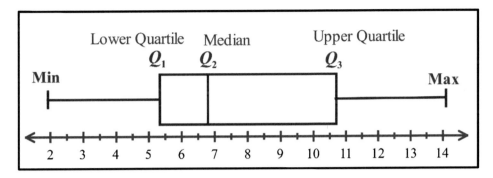

▨ The ends of the box are the **upper** (Q_3) and **lower** (Q_1) quartiles, so the box spans the **interquartile range** (IQR)

$$IQR = Q_3 - Q_1$$

▨ The **median** (Q_2) is marked by a vertical line inside the box.
▨ The **whiskers** are the two lines outside the box that extend to the maximum and minimum observations.
▨ An **outlier** is a point which falls more than 1.5 times the interquartile range (1.5× IQR) above the third quartile or below the first quartile.

Important: A GDC may be used to produce **histograms** and **box-and-whisker plots**.

- If we **add** or **subtract** a positive constant value **c** **to/from** all the numbers in the dataset, the **mean** (and the **median**) will also **increase** or **decrease**, respectively, by **c** and the **standard deviation** (and the **variance**) **will** remain the **same**.

- If we **multiply** or **divide** all the numbers in the dataset by a positive constant value **c**, the **mean**, the **median** and the **standard deviation** will be **multiplied** or **divided**, respectively, by **c**. It follows that the **variance** will be **multiplied** or **divided** by **c^2**.

Example

A data set has a mean of 24 and a standard deviation of 4.

(a) Each value in the data set has 7 added to it. Find the new mean, the new standard deviation, and the new variance.

(b) Each value in the data set is multiplied by 5. Find the new mean, the new standard deviation, and the new variance.

Answer

(a) The new mean is $24 + 7 = 31$, and both the standard deviation and variance remain the same.

(b) The new mean is $24 \times 5 = 120$, the new standard deviation is $4 \times 5 = 20$ and the new variance is $20^2 = 400$.

Example of cumulative frequency diagram

The following cumulative frequency table displays the marks obtained in a test by a group of 80 students. The cumulative frequency is calculated by accumulating the frequencies as we move down the table.

Grades	Frequency	Cumulative frequency
[0-20]	5	5
(20-40]	10	15
(40-60]	25	40
(60-80]	25	65
(80-100]	15	80

The corresponding cumulative frequency diagram is presented below:

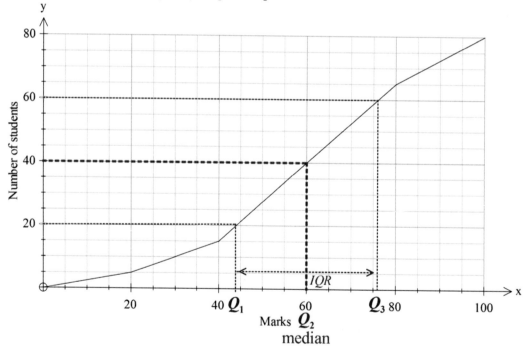

The **median** is estimated using the 50th percentile. As 50% of 80 is 40, we draw a horizontal line parallel to x-axis (Marks), passing through 40 until this line cuts the curve. Then we draw a vertical line parallel to y-axis (Number of students) until it reaches the x-axis, in this case, the value is 60 marks. Similarly, the lower and upper quartiles Q_1, Q_3 can be found at a cumulative frequency of 20 (25th percentile) and 60 (75th percentile) number of students respectively. Following the above steps, we get $Q_1 = 44$ and $Q_3 = 76$. Thus, the interquartile range is $IQR = Q_3 - Q_1$.

Probability

- A **sample space** U (Universal Set) is the set of all possible outcomes of an experiment.

- An **event** is one or more outcomes of an experiment. Mathematically, an event is a subset of a sample space. For example, scoring a four on the throw of a die.

- An **event** is **simple** if it consists of just a single outcome, and is **compound** otherwise.

- A sample space is discrete if it consists of a finite or countable infinite set of outcomes. A sample space is continuous if it contains an interval (either finite or infinite) of real numbers.

- The set containing no elements is called an **empty set** (or null set) and denote it by $\emptyset$.

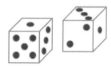

Example

If we toss a coin three times and record the result, the sample space is

$$U = \{HHH, HHT, HTH, HTT, THH, THT, TTH, TTT\}$$

where (for example) THH means 'Tails on the first toss, then heads, then heads again.'

The **theoretical probability** of an occurring event A is given by

$$P(A) = \frac{N(A)}{N(U)} = \frac{Number\ of\ outcomes\ in\ which\ A\ occurs}{Total\ number\ of\ outcomes\ in\ the\ sample\ space\ U}$$

Axioms of probability

1. $0 \leq P(A) \leq 1$

2. $P(\emptyset) = 0$ and $P(U) = 1$

3. If A and B are both subsets of U and they are **mutually exclusive** ($A \cap B = \emptyset$), then

$$P(A \cup B) = P(A) + P(B)$$

Also, we have the following propositions:

- $P(A') = 1 - P(A)$, where A' is the complement event of A
- If $A \subseteq B$ then $P(A) \leq P(B)$
- $P(A \cup B) = P(A) + P(B) - P(A \cap B)$

Example

A six-sided die is rolled twice. What is the probability that the sum of the numbers is at least 10?

Answer

The number of elements in the sample space is $6^2 = 36$. To obtain a sum of 10 or more, the possibilities for the numbers are $(4,6), (5,5), (6,4), (5,6), (6,5)$ or $(6,6)$. So the probability of the event "that the sum of the numbers is at least 10" is $\dfrac{6}{36} = \dfrac{1}{6}$.

Example

A box contains 4 red and 5 blue disks. A disk is randomly selected and has its color noted. The disk is not replaced and a second disk is then selected.

(a) Find the probability that the disks will be of a different color.
(b) Find the probability that they will be both red.

Answer

The tree diagram for this information is:

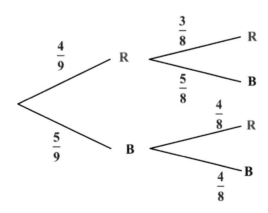

(a) $P(different\ color) = P(RB) + P(BR) = \dfrac{4}{9} \times \dfrac{5}{8} + \dfrac{5}{9} \times \dfrac{4}{8} = \dfrac{40}{72}$

(b) $P(both\ red) = P(RR) = \dfrac{4}{9} \times \dfrac{3}{8} = \dfrac{12}{72}$

Venn Diagrams

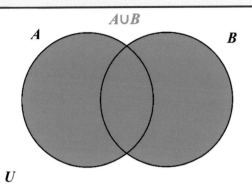

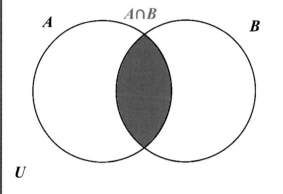

The **Union** of A and B is the set whose elements are all the elements of A and B.

$$A \cup B = \{x : x \in A \ or \ x \in B\}$$

$$P(A \cup B) = P(A) + P(B) - P(A \cap B)$$

The **Intersection** of A and B is the set consisting of the elements that belong to both A and B.

$$A \cap B = \{x : x \in A \ and \ x \in B\}$$

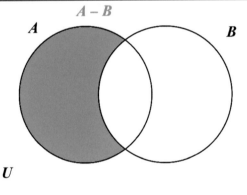

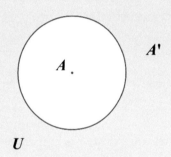

The **Difference** of A and B is the set consisting of those elements of A that are not in B.

$$A - B = \{x : x \in A \ and \ x \notin B\}$$

$$P(A - B) = P(A) - P(A \cap B)$$

The **Complement** of A is the set consisting of those elements of U that are not in A.

$$A' = A^c = U - A = \{x \in U : \ x \notin A\}$$

$$P(A') = 1 - P(A)$$

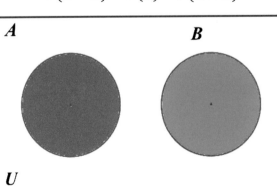

If $A \cap B = \emptyset$ the events A and B are said to be **disjoint** or **mutually exclusive**. Disjoint sets are sets which do not have elements in common.

$$P(A \cup B) = P(A) + P(B)$$

Conditional Probability

The probability of an event given that another event has already occurred is a conditional probability.

If A and B are two events, then the **conditional probability** of event A **given (/)** an event B can be found by the formula

$$P(A/B) = \frac{P(A \cap B)}{P(B)}, P(B) \neq 0$$

It should also be noted that usually $P(A/B) \neq P(B/A)$

Example

A lot of 1000 semiconductor chips contains 40 that are defective. Two are selected randomly, without replacement, from the lot.

(a) What is the probability that the first one selected is defective?

(b) What is the probability that the second one selected is defective **given** that the first one was defective?

(c) What is the probability that both are defective?

Solution

Let A: The first one selected is defective

and B: The second one selected is defective

Then:

(a) $P(A) = \frac{40}{1000} = 0.04$

(b) $P(B/A) = \frac{P(B \cap A)}{P(A)} = \frac{\frac{40}{1000} \cdot \frac{39}{999}}{\frac{40}{1000}} = \frac{39}{999}$

(c) $P(B \cap A) = \frac{40}{1000} \cdot \frac{39}{999} = \frac{156}{99900}$

Independence

Two events A and B are said to be **independent** if

$$P(A \cap B) = P(A) \times P(B)$$

or $P(A/B) = P(A)$ and $P(B/A) = P(B)$

Properties of independence

- If A and B are independent then A and B' are independent.
- If A and B are independent, so are A' and B'.

Note: A common mistake is to confuse whether two events are **independent** or **mutually exclusive**. A and B are mutually exclusive events or disjoint if $P(A \cap B) = 0$, that is, the occurrence of one precludes that of the other.

Examples

1. If $P(A/B) = 0.5$, $P(B) = 0.7$ and $P(A) = 0.4$, are the events A and B independent?

Answer The events are not independent because $P(A/B) \neq P(A)$

2. Given that $P(A) = 0.7$, $P(B) = 0.5$ and that A and B are **independent** events. Find the probability of the following events: **(a)** $A \cap B$ **(b)** $A \cup B$ **(c)** A/B' **(d)** $A' \cap B$

   Solution The events are independent, therefore

(a) $P(A \cap B) = P(A) \times P(B) = 0.7 \times 0.5 = 0.35$

(b) $P(A \cup B) = P(A) + P(B) - P(A \cap B) = 0.7 + 0.5 - 0.35 = 0.85$

(c) $P(A/B') = \dfrac{P(A \cap B')}{P(B')} = \dfrac{P(A) - P(A \cap B)}{1 - P(B)} = \dfrac{0.7 - 0.35}{1 - 0.5} = \dfrac{0.35}{0.5} = 0.7$

or we know that "If A and B are independent then A and B' are also independent",

therefore $P(A/B') = P(A) = 0.7$

(d) $P(A' \cap B) = P(A') \times P(B) = 0.3 \times 0.5 = 0.15$

Discrete Probability Distributions

A **random variable** is called **discrete** if it has either a finite or a countable number of possible values.

A **discrete probability distribution** describes the probability of occurrence of each value of a discrete random variable. The related function that outputs the probabilities of the respective values of the discrete random variable can assume, is called **probability density function (pdf)**.

If X is a **discrete random variable** with $P(X = x_i)$, $i = 1,2, \ldots n$ then

1. $0 \leq P(X = x_i) \leq 1$ for all values of x_i

2. $\sum_{i=1}^{n} P(X = x_i) = P(X = x_1) + \ldots + P(X = x_n) = 1$

3. The **expectation** of the random variable X is

$$E(X) = \mu = \sum_{i=1}^{n} x_i P(X = x_i)$$

4. The **variance** is defined by the following formula

$$Var(X) = E(X^2) - [E(X)]^2$$

Note: If we have a **fair game** then $E(X) = 0$ where X represents the gain of one of the players.

- For a discrete random variable X, the **mode** is any value of x with the highest probability and it may not be unique.

Example

The following table shows the probability distribution of a discrete random variable X.

x	0	2	4
$P(X = x)$	0.2	$3m$	m

a) Find the value of m
b) Find the expected value of X.

Solution

a) $\sum_{i=1}^{3} P(X = x_i) = 0.2 + 3m + m = 4m + 0.2 \xrightarrow{\text{the sum of probabilities equals 1}} 4m + 0.2 = 1$

$\Rightarrow 4m = 0.8 \Rightarrow m = 0.2$

b) $E(X) = \sum_{i=1}^{3} x_i P(X = x_i) = 0 \times 0.2 + 2 \times (3 \times 0.2) + 4 \times (0.2) = 2$

Binomial Distribution

The **Binomial distribution** is a discrete probability distribution. It describes the outcome of n **independent** trials. Each trial is assumed to have only two outcomes, either **success** or **failure**. The probability of a success, denoted by p, remains **constant** from trial to trial. The probability of having exactly r **successes** in n independent trials is given by the following formula:

$$P(X = r) = \binom{n}{r} p^r (1 - p)^{n-r}$$

$$\text{where } \binom{n}{r} = \frac{n!}{r!(n-r)!} \text{ and } r = 0, 1, \dots n$$

If a discrete random variable X follows a Binomial distribution with parameters n and p, which is written $X \sim B(n, p)$, then it has

Expected value (mean)	Variance
$E(X) = np$	$Var(X) = np(1 - p)$

Example

A **fair** coin is tossed **seven** times. Calculate the probability of obtaining:

(i) Exactly four heads.

(ii) At least two heads.

Solution

(i) Let X denote the number of heads, $X \sim B(7, 0.5)$

$$P(X = 4) = \binom{7}{4}(0.5)^4(0.5)^{7-4} = 0.273 \ (3 \text{ s.f.})$$

(ii) Let X denote the number of heads, $X \sim B(7, 0.5)$

$$P(X \geq 2) = 1 - P(X \leq 1) = 1 - \big(P(X = 0) + P(X = 1)\big) = 0.938 (3 \text{ s.f.})$$

where $P(X = 0) = \binom{7}{0}(0.5)^0(0.5)^{7-0} = 0.00781 (3 \text{ s.f.})$

$$P(X = 1) = \binom{7}{1}(0.5)^1(0.5)^{7-1} = 0.0546875$$

TI 84 + (Example)		Casio fx9860 series, fx-CG20, fx-CG50 (Example)	
$X \sim B(7,0.5)$ $P(X = 4)$	**binompdf**($numtrials, p, x$) Computes a probability at x for the discrete binomial distribution with the specified *numtrials* and probability p of success on each trial. 2nd [DISTR] **DISTR** **A:binompdf(7,0.5,4)**	$X \sim B(7,0.5)$ $P(X = 4)$	On the initial STAT mode screen →F5(DIST) →F5(BINM) →F1(Bpd) we set Data: **Variable**, x: **4**, Numtrial: **7**, p: **0.5** →Execute $P(X = 4) = 0.273 \, (3 \, \text{s.f.})$
$X \sim B(7,0.5)$ $P(X \le 1)$	**binomcdf**($numtrials, p, x$) Computes a cumulative probability at x for the discrete binomial distribution with the specified *numtrials* and probability p of success on each trial. 2nd [DISTR] **DISTR** **B:binomcdf(7,0.5,1)**	$X \sim B(7,0.5)$ $P(X \le 1)$	On the initial STAT mode screen →F5(DIST) →F5(BINM) →F2(Bcd) we set Data: **Variable**, x: **1**, Numtrial: **7**, p: **0.5** →Execute $P(X \le 1) = 0.063$ and then $P(X \ge 2) = 1 - P(X \le 1) = 0.938(3 \, \text{s.f.})$

Note: Casio models **fx-CG20, fx-CG50** calculate directly the probability $P(X \ge 2)$.

Normal Distribution

The **normal distribution** is a theoretical ideal distribution. Real-life empirical distributions never match this model perfectly. However, many things in life do approximate the normal distribution, and are said to be "normally distributed".

The normal distribution has the following properties:

▨ Its shape is symmetric about the **mean (μ)**, which is also the **median** and the **mode** of the distribution.

▨ It is a bell-shaped curve, with tails going down and out to the left and right and the x-axis is a horizontal asymptote.

▨ Its standard deviation (σ), measures the distance on the distribution from the mean to the inflection point.

▨ The total area under the curve is equal to 1.

▨ Approximately the 68 percent of its values lie within one standard deviation of the mean.

▨ Approximately the 95 percent of its values lie within two standard deviations of the mean.

▨ Approximately the 99.7 percent of them lies within three standard deviations of the mean.

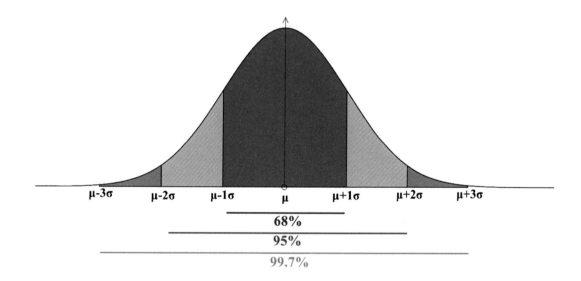

Example

The weights of a group of men are normally distributed with a mean of 80 kg and a standard deviation of 15 kg.

(i) A man is chosen at random. Find the probability that the man's weight is greater than 90 kg.

(ii) In this group, 20% of men weigh less than w kg. Find the value of w.

Solution

The required probability for **(i)** is represented in the following diagram.

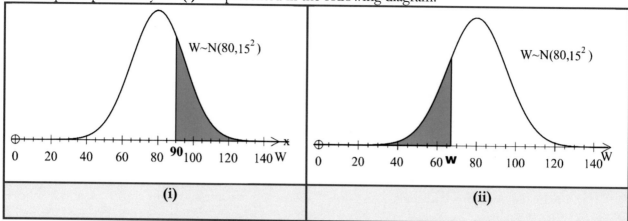

(i)	**(ii)**

To find this probability, we can use GDC.

▣ For **Casio fx9860 series, fx-CG20, fx-CG50**, we perform the following commands:

On the initial STAT mode screen →F5(DIST) →F1(NORM) →F2(Ncd)

we set Lower: **90**, Upper: **10^99**, σ: **15**, μ: **80** → Execute

$$P(W > 90) = 0.252 \ (3 \text{ s. f.})$$

▣ For **TI 84 +**

2nd →VARS →2 →Enter four parameters (lower limit: **90**, upper limit: **10^99**, mean: **80**, standard deviation: **15**)

(ii) To find the value of w, we have to use the inverse normal which gives us an x-value if we input the area (probability region) to the left of the x-value.

To find this probability, we can use GDC.

▣ For **Casio fx9860 series, fx-CG20, fx-CG50**

On the initial STAT mode screen →F5(DIST) →F1(NORM) →F3(InvN)

we set Tail: Left, Area(probability less than): **0.20**, σ: **15**, μ: **80** →Execute

▣ For **TI 84 +**

2nd →VARS →3 →Enter three parameters (probability less than - area: **90**, mean: **80**, standard deviation: **15**)

$$P(W < w) = 0.20 \Rightarrow w = 67.4$$

Important: There are many normal distribution exercises where we have to find the **mean**(μ) and/or the **standard deviation** (σ). In this case, we have to convert this normal distribution $X \sim N(\mu, \sigma^2)$ to a **standard normal distribution** $Z \sim N(0, 1^2)$ by using the formula $z = \dfrac{x - \mu}{\sigma}$.

Correlation - Regression

Correlation

Correlation describes the degree of relationship between two variables.

The most common measure of correlation in statistics is the Pearson correlation coefficient. It shows the linear relationship between two sets of data. It's represented by r and it has a range of

$$-1 \leq r \leq 1$$

The sign of r indicates the direction of the correlation.

$r > 0$ means a **positive** correlation

$r < 0$ means a **negative** correlation

The magnitude of r indicates the strength of the correlation.

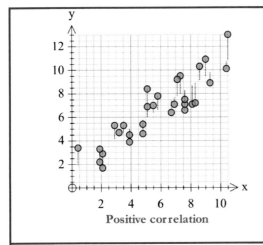 Positive correlation	A positive correlation is a relationship between two variables where if one variable increases, the other one also increases. If the independent variable (x) increases then the dependent variable (y) also increases. $r > 0$
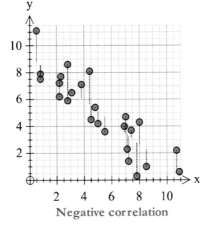 Negative correlation	A negative correlation means that there is an inverse relationship between these two variables. When one variable decreases, the other increases. If the independent variable (x) increases then the dependent variable (y) decreases. $r < 0$

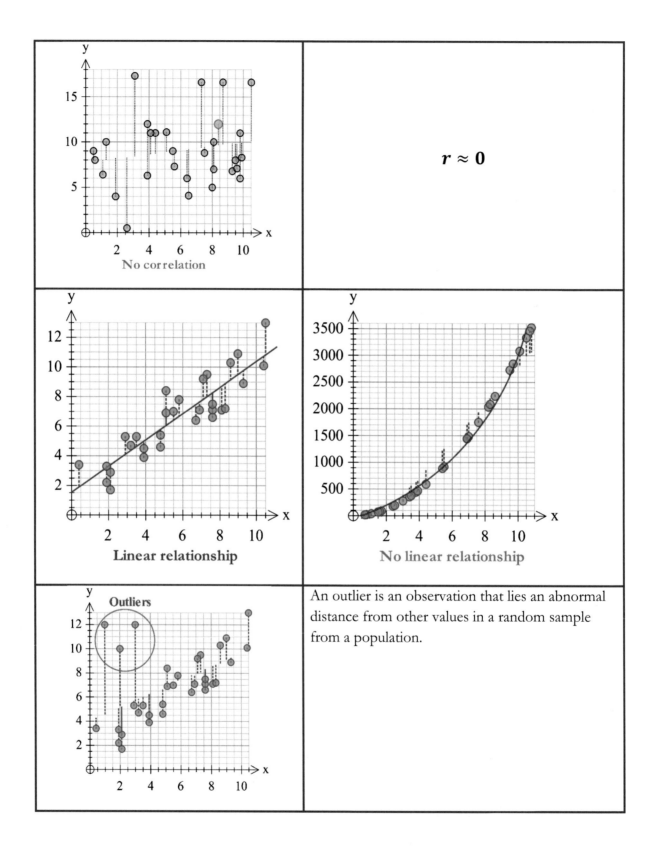

$$r \approx 0$$

No correlation

Linear relationship

No linear relationship

Outliers

An outlier is an observation that lies an abnormal distance from other values in a random sample from a population.

Positive correlation		Negative correlation	
$r = 1$	perfect positive	$r = -1$	perfect negative
$0.95 \leq r < 1$	very strong positive	$-1 \leq r < -0.95$	very strong negative
$0.75 \leq r < 0.95$	strong positive	$-0.95 \leq r < -0.75$	strong negative
$0.50 \leq r < 0.75$	moderate positive	$-0.75 \leq r < -0.50$	moderate negative
$0.10 \leq r < 0.50$	weak positive	$-0.50 \leq r < -0.10$	weak negative
$0 \leq r < 0.10$	no correlation	$-0.10 < r \leq 0$	no correlation

Line of best fit

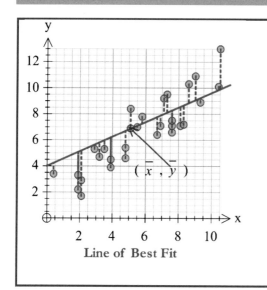

Line of Best Fit

A line of best fit is a straight line of the form

$$y = ax + b$$

that is the best approximation of the given set of data.

A line of best fit can be roughly determined by drawing a straight line on a scatter plot so that the number of points above the line and below the line is about equal, the line passes through as many points as possible and passes through the mean point $(\bar{x}, \bar{y})$.

Note: The more the correlation coefficient (r) is closer to 1 or -1 , the more accurate the predictions, produced by the line of best fit, are.

Interpolation

We could use the **line of best fit** to predict the value of the dependent variable (y) for an independent variable (x) that is **within the range** of our data. In this case, we are performing **interpolation**.

Extrapolation

We could use the **line of best fit** to predict the value of the dependent variable for an independent variable that is **outside the range** of our data. In this case, we are performing **extrapolation**.

Of the two methods, **interpolation** is preferred. This is because we have a greater likelihood of obtaining a valid estimate than with **extrapolation**.

Vectors

A **vector** is an object that has both a **magnitude** and a **direction**.

Magnitude is defined as the length of a vector

$$|a| = \left|\begin{pmatrix} a_1 \\ a_2 \end{pmatrix}\right| = \sqrt{a_1{}^2 + a_2{}^2} \quad \text{(2D)}$$

$$|a| = \left|\begin{pmatrix} a_1 \\ a_2 \\ a_3 \end{pmatrix}\right| = \sqrt{a_1{}^2 + a_2{}^2 + a_3{}^2} \quad \text{(3D)}$$

The zero vector **0** has zero magnitude, $|\mathbf{0}| = 0$ and has no definite direction.

Two vectors are **equal** if they have the same magnitude and direction.

Two vectors are **parallel** if they have the same direction or are in exactly opposite directions.

Two vectors are **parallel** if one is a scalar multiple of the other ($k \epsilon \mathbb{R}$).

$$a \parallel b \Leftrightarrow a = kb$$

A **unit vector** is a vector that has a magnitude of 1.

Any vector a can become a **unit vector** $\hat{a}$ by dividing it by its magnitude, i.e. $\hat{a} = \dfrac{a}{|a|}$.

The unit vectors $\boldsymbol{i}, \boldsymbol{j}$ and $\boldsymbol{k}$ are also known as the **base unit vectors** of x, y and z-axis respectively.

$$i = \begin{pmatrix} 1 \\ 0 \\ 0 \end{pmatrix}, j = \begin{pmatrix} 0 \\ 1 \\ 0 \end{pmatrix}, k = \begin{pmatrix} 0 \\ 0 \\ 1 \end{pmatrix}$$

For example, the vector $a = \begin{pmatrix} -2 \\ 5 \\ 4 \end{pmatrix}$ can also be written as $a = -2\boldsymbol{i} + 5\boldsymbol{j} + 4\boldsymbol{k}$

Operations on Vectors

Vector Addition

The sum of two vectors a and b is defined as the diagonal of the parallelogram formed when the two vectors a and b are placed at the same point, as is described in the following diagram.

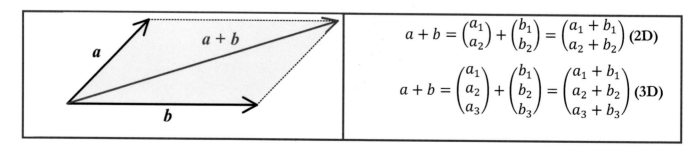

$$a + b = \begin{pmatrix} a_1 \\ a_2 \end{pmatrix} + \begin{pmatrix} b_1 \\ b_2 \end{pmatrix} = \begin{pmatrix} a_1 + b_1 \\ a_2 + b_2 \end{pmatrix} \quad \text{(2D)}$$

$$a + b = \begin{pmatrix} a_1 \\ a_2 \\ a_3 \end{pmatrix} + \begin{pmatrix} b_1 \\ b_2 \\ b_3 \end{pmatrix} = \begin{pmatrix} a_1 + b_1 \\ a_2 + b_2 \\ a_3 + b_3 \end{pmatrix} \quad \text{(3D)}$$

The negative of a vector

The vector $-a$ is defined to be the same as vector a, but with the opposite direction and $a + (-a) = 0$

The negative of $\overrightarrow{AB}$ is $-\overrightarrow{AB}$ or $\overrightarrow{BA}$.

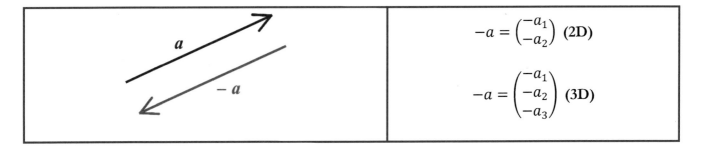

$$-a = \begin{pmatrix} -a_1 \\ -a_2 \end{pmatrix} \text{ (2D)}$$

$$-a = \begin{pmatrix} -a_1 \\ -a_2 \\ -a_3 \end{pmatrix} \text{ (3D)}$$

Vector subtraction

Vector subtraction $a - b$ is defined as:

$a + (-b)$, that is, adding the negative of the subtracted vector b.

$$a - b = \begin{pmatrix} a_1 \\ a_2 \end{pmatrix} - \begin{pmatrix} b_1 \\ b_2 \end{pmatrix} = \begin{pmatrix} a_1 - b_1 \\ a_2 - b_2 \end{pmatrix} \text{ (2D)}$$

$$a - b = \begin{pmatrix} a_1 \\ a_2 \\ a_3 \end{pmatrix} - \begin{pmatrix} b_1 \\ b_2 \\ b_3 \end{pmatrix} = \begin{pmatrix} a_1 - b_1 \\ a_2 - b_2 \\ a_3 - b_3 \end{pmatrix} \text{ (3D)}$$

The vector between two points

If A, B are two points with position vectors $\overrightarrow{OA} = \begin{pmatrix} a_1 \\ a_2 \\ a_3 \end{pmatrix}, \overrightarrow{OB} = \begin{pmatrix} b_1 \\ b_2 \\ b_3 \end{pmatrix}$ respectively, then the position vector of

B relative to A is $\overrightarrow{AB} = \overrightarrow{OB} - \overrightarrow{OA} = \begin{pmatrix} b_1 \\ b_2 \\ b_3 \end{pmatrix} - \begin{pmatrix} a_1 \\ a_2 \\ a_3 \end{pmatrix} = \begin{pmatrix} b_1 - a_1 \\ b_2 - a_2 \\ b_3 - a_3 \end{pmatrix}$

Multiplication by a scalar

Multiplying a vector a by a scalar k gives a new vector with the same direction but with a magnitude which is multiplied by this scalar.

$$ka = k \begin{pmatrix} a_1 \\ a_2 \\ a_3 \end{pmatrix} = \begin{pmatrix} ka_1 \\ ka_2 \\ ka_3 \end{pmatrix}$$

Scalar Product

The **scalar product** or **dot product** can be defined for two vectors $\boldsymbol{a}$ and $\boldsymbol{b}$ by

$$a \cdot b = |a||b|cos\theta$$

$$cos\theta = \frac{a \cdot b}{|a||b|}$$

where θ is the **angle** between the vectors and $|a|, |b|$ are the **magnitudes** (lengths) of each vector.

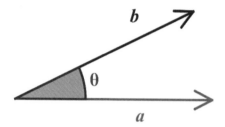

If the vectors are expressed in component form, the **dot product** can also be expressed in the following forms:

$\begin{pmatrix} a_1 \\ a_2 \end{pmatrix} \cdot \begin{pmatrix} b_1 \\ b_2 \end{pmatrix} = a_1b_1 + a_2b_2$ **(2D)**	$\begin{pmatrix} a_1 \\ a_2 \\ a_3 \end{pmatrix} \cdot \begin{pmatrix} b_1 \\ b_2 \\ b_3 \end{pmatrix} = a_1b_1 + a_2b_2 + a_3b_3$ **(3D)**

Properties of Scalar Product

If a, b, c are vectors and k is a scalar then

1. $a \cdot b = b \cdot a$

2. $a \cdot a = |a|^2$

3. $a \cdot (b + c) = a \cdot b + a \cdot c$

4. $(ka) \cdot b = k(a \cdot b) = a \cdot (kb)$

5. $0 \cdot a = 0$

▫ If two vectors are **perpendicular** then,

$$a \cdot b = 0$$

▫ If two vectors are **parallel** then,

$$a \cdot b = |a||b|cos0^o = |a||b|$$

or

$$a \cdot b = |a||b|cos180^o = -|a||b|$$

Lines

Vector equation of a line

$$r = a + \lambda b$$

where a is a **position vector** (a point which lies on the line), b the **direction vector** (a vector parallel to the line) and $\lambda \in \mathbb{R}$ is called the **parameter**.

Parametric form of the equation of a line

$$\begin{aligned} x &= a_1 + \lambda b_1 \\ y &= a_2 + \lambda b_2 \\ z &= a_3 + \lambda b_3 \end{aligned}$$

where $a = \begin{pmatrix} a_1 \\ a_2 \\ a_3 \end{pmatrix}$ is a **position vector**, $b = \begin{pmatrix} b_1 \\ b_2 \\ b_3 \end{pmatrix}$ the **direction vector** of the line and $\lambda \in \mathbb{R}$ is called the **parameter**.

Cartesian equations of a line

$$\frac{x - a_1}{b_1} = \frac{y - a_2}{b_2} = \frac{z - a_3}{b_3}$$

where $a = \begin{pmatrix} a_1 \\ a_2 \\ a_3 \end{pmatrix}$ is a position vector and $b = \begin{pmatrix} b_1 \\ b_2 \\ b_3 \end{pmatrix}$ the direction vector of the line.

Note: The acute angle θ between two lines l_1, l_2 can be found by using the following formula

$$cos\theta = \frac{|b \cdot c|}{|b||c|}$$

where b, c are the direction vectors of the lines l_1 and l_2 respectively.

Relative Positions of two Lines

In 2-D or 3-D

- **Intersecting lines** meet at a unique point.

- **Perpendicular lines** have their direction vectors perpendicular, i.e. their dot product equals zero.

- **Parallel lines** are lines that never touch. Their direction vectors are parallel (scalar multiples of one another).

- **Skew lines** (only in 3D) are lines that do not intersect and are not parallel.

- **Coincident (identical) lines** are lines that lie exactly on top of each other. These lines are parallel and meet at a point. So, they are the same line.

Example

Find the point of intersection between the lines defined by the following two equations:

$$L_1:r = \begin{pmatrix} 1 \\ 0 \\ 3 \end{pmatrix} + \lambda \begin{pmatrix} -1 \\ 1 \\ -2 \end{pmatrix} \qquad\qquad L_2:r = \begin{pmatrix} 1 \\ -1 \\ 4 \end{pmatrix} + \mu \begin{pmatrix} 2 \\ -1 \\ 3 \end{pmatrix}$$

The parametric form of these lines is:

$$L_1: x = 1 - \lambda, \qquad y = \lambda, \qquad z = 3 - 2\lambda$$

$$L_2: x = 1 + 2\mu, \qquad y = -1 - \mu, \qquad z = 4 + 3\mu$$

For the lines to intersect there must be a value of λ and μ that will provide the same point lying on both L_1 and L_2. Using the parametric equations we equate the coordinates and try to determine the common point.

$$1 - \lambda = 1 + 2\mu \quad (1)$$
$$\lambda = -1 - \mu \quad (2)$$
$$3 - 2\lambda = 4 + 3\mu \quad (3)$$

Solving for λ and μ using (1) and (2) we obtain: $\lambda = -2$ and $\mu = 1$.

Checking these values in (3), we have $3 - 2(-2) = 7 = 4 + 3(1)$

Using $\lambda = -2$ or $\mu = 1$,

$$x = 1 - (-2) = \mathbf{3}, \qquad y = -2, \qquad z = 3 - 2(-2) = \mathbf{7}$$

we can determine the point of intersection $(3, -2, 7)$.

Kinematics

An important application of vectors is on kinematics.

The vector equation

$$r = a + tb$$

represents the position of a particle at **time** t, where a is its **initial position** and b is its **velocity vector**. The **speed** of the particle is the magnitude of the **velocity vector** $|b|$.

Example

Let $r = \binom{1}{2} + t\binom{5}{6}$ the vector equation of the motion of a particle, where t represents the time in seconds and distances are measured in meters.

(a)   Find the particle's initial position.

(b)   Find the velocity vector of the particle.

(c)   Find the speed of the particle.

Solution

(a)   The initial position is when $t = 0$, so $r = \binom{1}{2}$.

(b)   The velocity vector is $\binom{5}{6}$.

(c)   Speed (s) is the magnitude of velocity, thus $s = \left|\binom{5}{6}\right| = \sqrt{5^2 + 6^2} = \sqrt{61}\ ms^{-1}$.

References

1. Mathematics SL formula booklet, First examinations 2014, Diploma Programme, International Baccalaureate®.

2. Mathematics SL guide, First examinations 2014, Diploma Programme, International Baccalaureate®.

3. TI-84 Plus and TI-84 Plus Silver Edition Guidebook.

Made in the USA
Middletown, DE
07 April 2019